Business Sustainability Model for Malaysian Housing Developers

Mastura Jaafar • Azlan Raofuddin Nuruddin
Syed Putra Syed Abu Bakar

Business Sustainability Model for Malaysian Housing Developers

Springer

Mastura Jaafar
School of Housing, Building and Planning
Universiti Sains Malaysia
Penang, Malaysia

Azlan Raofuddin Nuruddin
School of Housing, Building and Planning
Universiti Sains Malaysia
Penang, Malaysia

Syed Putra Syed Abu Bakar
School of Housing, Building and Planning
Universiti Sains Malaysia
Penang, Malaysia

ISBN 978-981-10-5265-1 ISBN 978-981-10-5266-8 (eBook)
DOI 10.1007/978-981-10-5266-8

Library of Congress Control Number: 2017945272

Printed on acid-free paper

This Springer imprint is published by Springer Nature
The registered company is Springer Nature Singapore Pte Ltd.
The registered company address is: 152 Beach Road, #21-01/04 Gateway East, Singapore 189721, Singapore

Foreword by the Vice-Chancellor of Universiti Sains Malaysia

Allow me to begin by congratulating the authors involved in extending the idea of sustainability to the viability of a business model in the housing sector. Sustainability has always been the core idea at Universiti Sains Malaysia, as we formulate and implement policies and research outcomes for the betterment of all.

Property development remains a booming business sector in the country. It offers the populace the availability of choices and the opportunity to become more established on a personal level. At the same time, the industry promotes growth in the skills-based job market too. However, as with any business model, the certainty in continuity is a factor that could sustain businesses in the long run, as they face the ever-growing and challenging market. Therefore, the publication of this book, with the support of a university grant, is a truly welcomed effort, as it could serve as a guide or reference to those involved in the industry.

I applaud this book also not only for its timeliness but for putting a 'face' behind the development firms, for 'humanising' the entity behind the realisation of the all too basic of human needs. Much maligned and abused, and deservingly so at times, this effort dismantles much of the myth of the 'snake oil' peddlers of properties. We are informed of anecdotes that illustrate the difficulties, obstacles and a retinue of impediments that developers have to confront. At the same time, we are treated to first-hand accounts from property players of the risks, challenges and struggles they had to endure. We are then regaled about tales of their fortitude, perseverance and sheer industry. They do all this, and more, not just for the pursuit of profit but to enable the current and future generations to have the proverbial 'roof over their heads'.

I also welcome the manner in which this subject has been treated, as it covers a fairly comprehensive range of small- to medium-sized firms right up to major public-listed conglomerates. Interestingly still, certain core values were painstakingly extracted and compared, making the book an extremely useful reference point for potential developers, besides providing current ones with clear-cut examples of learning from the country's best. While some of the issues appear mundane and obvious, they still needed to be stated unequivocally and convincingly – this effort has managed to successfully capture and convey both.

Vice-Chancellor
Universiti Sains Malaysia

Professor Datuk Dr. Asma Ismail, FASc

Preface

Housing is an indispensable need for human beings. It not only reflects a certain standard of living, but its manifestation has transcended into a symbol of modernity and lifestyle. In Malaysia, the Real Estate and Housing Developers' Association (REHDA) estimates that about 150,000 units of houses are required annually to fulfil the population demands and household growth. To accommodate the growing demand, the government of Malaysia has been encouraging the involvement of private housing developers since the 1970s. The role of private residential developers is significantly increasing and is considered a particularly important component of the industry for national economic development.

Pragmatically, a housing developer is a coordinator of housing production that works through the accumulation of necessary resources and capabilities and to progressively monitor the residential project until its final delivery. Within the development process, the housing developer deals with numerous land and construction-related activities such as site identification and assembly, subdivision approval, site supervision and marketing the project. In Malaysia, housing developers normally comprise private developers, state-owned enterprises and cooperative societies and are involved in all categories of housing provision encompassing high-, medium- and low-cost houses as defined under the Housing Developers' Act (Control and Licensing) 1996. To develop and sell more than four houses, residential developers must obtain a licence from the Ministry of Urban Wellbeing, Housing and Local Government (MUWHLG) and are encouraged to register as members with REHDA.

Basically, housing development involves a huge amount of money and entails diverse and complex processes such as land subdivision, building construction, final product delivery, advertising, sales and marketing. In today's globally competitive environment, with twin facets of rapid technological progress and high customer expectations, companies find it harder to 'beat' the competition by depending only on their own capacity. For the real estate sector, notwithstanding the high capital investments, the composition of industry players is rather 'kaleidoscopic'; those who are unable to perform will be forced to exit the industry, only to be replaced with fledglings who may lack experience but are hungry for success for which the sector is well known for. Without sufficient industry knowledge, however, these new entrants will invariably be exposed to the real possibility of business failure, and the cycle is then repeated. The compelling question therefore is how new housing developers can enhance their competitive advantage when entering the real estate market and be able to sustain within the context of such a profitable but volatile industry.

Simply put, housing developers gain profits from the margins derived from house selling prices and, to an extent, a sense of self-satisfaction from the sale of the products, which oftentimes represents the single largest investment people will ever make. Developers are, by nature, commercially driven in a business fraught with risks. As such, the only real criterion they can ratchet up by themselves is sheer motivation, intrinsically the desire to become a successful entrepreneur and, in so doing, to transform their lives. Achieving competitive advantage depends on the ability of housing developers to mobilise and exploit the linkages of the numerous resources within their business environment. Each step in housing development necessitates a close and mutual relationship between owner-managers and firm capabilities, bearing in mind exogenous elements of the business such as government intervention, economic condition and others of the ilk, all of which are touted to affect the organisational performance.

This book focuses on three main themes – characteristics of owner-managers, firm resources and management and external factors – and will display 20 case studies of property developers in Malaysia, with each case study premised on the three themes. The case studies will be divided into three main sections comprising:

Small- and Medium-Scale Private Housing Developers
This section includes developers that are comparatively small to medium in size; engaged in development projects characterised by modest numbers on relatively small land banks, with average gross development value (GDV); and focused on specific locations. The bulk of Malaysian developers fall into this category, as are new market entrants, which may also include niche and boutique developers. However, players in this category may also constitute spin-offs from large organisations with non-property core businesses.

Large-Scale and Public-Listed Housing Developers
This group comprises large organisations characterised by big set-ups both in terms of personnel and locational presence, access to massive land banks and undertaking projects of high GDV, with the capacity to build large numbers and townships. The majority of developers in this category are public-listed entities with impressive historical track records, access to vast financial potential, award-winning achievements and outstanding reputation. While there are large development firms that are not public-listed, they normally have the carrying capacity to work with listed ones.

State-Owned Enterprises/Government-Linked Companies
As the name suggests, developers in this category are normally wholly owned entities of the federal or respective state governments. They can constitute small- and medium-sized firms but can also comprise large public-listed entities. While they

are commercially driven like most businesses, developers in this category have a duty towards fulfilling the agenda of homeownerships by the respective federal and state authorities; the main focus would appear to be on affordable housing in varied locations, although some developers have morphed to become aggressive niche players in their own right.

It is hoped that this book will provide some insightful ideas for the industry's newcomers and existing players in Malaysia. We would like to thank all the developers who had willingly participated in the interview sessions. For a record, this research work was supported by Universiti Sains Malaysia (Grant No: 1001/PPBGN/816202).

Penang, Malaysia

Mastura Jaafar
Azlan Raofuddin Nuruddin
Syed Putra Syed Abu Bakar

Acknowledgements

In the process of completing this book, we have benefited from several individuals' support. The authors would like to thank Wan Mohd Dzulkifli Wan Daud (of Maka Corporation Sdn Bhd) for his input and advice throughout the preparation process, including valuable comments and motivation to complete the whole journey of publication. Also, we must extend our hearty thanks to Muhd Sazlie Ibrahim, our former research officer, who worked tirelessly especially during the early process of manuscript writing. On top of that, we are indebted to all the research participants who had agreed to share their experience, to be documented in this book. Besides that, much appreciation goes out to the anonymous reviewers for taking their time to provide precious feedback and suggestions. Finally, we express our sincere gratitude to William Achauer, a Springer publishing editor, for his kind help and encouragement.

Contents

Notes

1. The term "Berhad", shortened to "Bhd", means "limited" while the term "Sendirian Berhad", shortened to "Sdn Bhd" or simply "S/B", means "private limited".
2. "RM" is the currency symbol for Malaysian Ringgit while "MYR" represents the currency code.
3. The Malay term "Taman" literally means park. However, in Malaysia, many residential areas were named with "Taman" in the beginning of denomination to denote residence or township. Other terms such as "Bandar", "Perumahan" and "Desa" were used to indicate similar attribute.
4. Bumiputera may best be defined as "son of the soil" in Malay language and refers to a Malaysian of indigenous origin such as Malay, Orang Asli, Kadazan--Dusun, Iban, Bidayuh or others of the ilk. Meanwhile, non-Bumiputera in Malaysia mainly represents Chinese and Indian.
5. In Peninsular Malaysia, Bumiputera contractors are graded from the lowest "F" Class (eligible for public contracts worth RM20,000 and below) to the highest "A" Class (eligible for public contracts worth RM10 million and above).
6. "Dato'" or "Datuk" is the royal honorary title, normally conferred by the hereditary ruler of the Malaysian states. Other notable titles in Malaysia include "Dato' Sri/Seri", "Datin Paduka", "Tan Sri" and "Tun".
7. UMNO is a Malay-based party and one of the three major component parties of the ruling coalition in Malaysia called the Barisan Nasional (BN) in Malay, or National Front in English. Two other political parties involved in the alliance are MCA (Malaysian Chinese Association) and MIC (Malaysian Indian Congress) that represent the Malaysian Chinese and Indian, respectively.
8. *Mudah.my* is an online marketplace that connects buyers and sellers in Malaysia by providing a platform for advertisement.
9. The noun phrase Menteri Besar (literally, "big minister") refers to the chief administrative official at the state level in Malaysia, follows basic Malay word order, with the adjective in post-nominal position. The role is equivalent to "Chief Minister".

Part I
Introduction: Private Housing Developers in Malaysia

Chapter 1
Private Housing Developers and Their Competitive Advantage

Keywords Business success • Competitive advantage • Resource-based view • Dynamic capability view • Firm performance

1.1 Introduction

Private housing developers usually will try to maximise their profit by quickly responding to market signals. Since they are entrepreneurs, they are naturally exposed to various types of risks. The approaches taken to tackle the risks against failure depend on the size of the firm and project. Due to ever-changing regulations, Abdul-Aziz, Ho, and Jaafar (2006) argue that private housing developers in Malaysia may resort to dubious approaches to minimise the risks and challenges.

Property development is an extremely creative process. Successful housing developers are a bit like film producers. They extend ideas, apply design and related approvals and seek financial capital and marketing mix, before starting the construction activities and then eventually making improvements to real property for sale or leasing. If they skip one of the said processes, they will confront with multiple risks when conducting development. Exposure to greater risks will lead their business to failure. Consequently, housing developers should be imbued with positive personal traits and tactful business plan. On top of that, the present scenario of the housing sector has transformed the strategies implemented by residential developers. Henceforth, the topic on success factors of housing developers seems opportune to be further unleashed.

1.2 Business Success and Competitive Advantage

Success remains elusive for businesspersons. A renowned economist, Michael Porter, considers success to be related with sustaining competitive advantage of the business that may lead to superior performance (Porter, 1985). The ability of an organisation to react and perform within the business realm through the identification of strengths and weaknesses over competitors will inextricably uplift the

M. Jaafar et al., *Business Sustainability Model for Malaysian Housing Developers*, DOI 10.1007/978-981-10-5266-8_1

competitiveness of the firms. Any such organisations that are able to outperform their peers will ultimately enjoy a competitive advantage and create a defensible position over fellow rivals. Paradoxically, a developer's success depends on the ability to synchronise the completion of a series of interconnected processes efficiently and at the correct time. On a similar note, long-term success in the heart of housing developers must be based on a set of interlinked success factors. Successful property developers are not born and they are definitely not created overnight. Since the business itself is part of entrepreneurial activities, business success can also be linked with the concept of entrepreneurship. Success can be gauged through many dimensions such as profitability, growth, market share and productivity or through nonfinancial measures such as satisfaction, self-fulfilment and competitiveness. A hybrid between tangible and intangible success measures could possibly reflect the true meaning of competitive advantage.

Specifically, one entrepreneurial question always evokes among related scholars and industry players: how can successful housing developers gain competitive advantage despite relentless challenges and obstacles? To address this question, two interwoven theories, i.e. resource-based view and dynamic capability view, were synthesised to analyse the competitive advantage of Malaysian successful housing developers in relation with the constant success. After the description on the research method, a number of extracted success factors from the selected Malaysian housing developers are adduced to holistically portray the subsequent chapters. Apart from self-capabilities and high-valued product, their success stems from a combination of attitude, experience, education, instinct and the lessons learnt from both victories and failures. External influences from government intervention, economic structure and sufficient funding capital are delineated as well.

1.3 The Underpinning Theories

A research work by Abdul-Aziz et al. (2006) towards Malaysian private residential developers has adopted the resource-based view (RBV) as its hypothetical basis to examine the valuable resources of housing development firms in relation to organisational competitiveness. The resource-based view (RBV) postulates that firms can obtain competitive advantage through the acquisition of resources that are rare, valuable and hard to imitate or replace (Barney, 1991). The resources comprise both tangible assets (e.g. land, capital and facilities) and intangible assets including capabilities, organisational processes, information and knowledge (Barney, 1991). According to RBV, firms in the same industry perform in a different way since they control dissimilar resources and capabilities. Those resources and capabilities can generate sustainable competitive advantage and subsequently prompt greater performance.

Teece, Pisano, and Shuen (1997) further expand the concept of RBV within rapidly volatile environments as firms need dynamic capabilities to enable resource alteration and competency reconfiguration to stay aggressive in business. Dynamic capability view (DCV) is developed over time through continuous practices that serves different purposes of resource integration and acquisition. Basically, it addresses internal and external competences in an attempt to adapt to the changing environment. In most cases, dynamic capabilities may be useful to enhance organisational abilities through knowledge creation and product development. The dynamic view of capabilities also highlights the strategic flexibility of resources and the alignment between the firm's assets which include the ability to sense and seize opportunities. Relatedly, success of housing developers depends on the firm's ability to constantly renew and reconfigure its resources and adjust them to overcome exogenous constraints.

Hence, the combination of both theories induces the establishment of integrated dynamic capabilities that mobilise and exploit business resources for further excellence in managing the business. The theory of integrated dynamic capabilities can also be applied to the field of residential development. In this domain, resources and capabilities are required by Malaysian housing developers to succeed, based on the capacity of both attributes in harnessing the opportunities and resolving the threats. In general, the verdicts of the theorists regarding competitive advantage have been utilised to identify reliable factors that drive firms or organisations towards business success and high performance.

Subsequently, the below discussion will succinctly describe the theoretical elements of business success and competitive advantage based on three emerging themes:

1.3.1 Theme 1: Characteristics of Owner-Managers

1.3.1.1 Demographic Profiles

For this research, demographic data such as level of education and prior experience seem to influence the success of Malaysian housing developers. On the other hand, demographic features like gender and ethnicity did not turn out to be significantly important in relation to business success, even though male entrepreneurs and Chinese builders seem to dominate the Malaysian property industry since its infancy stage. Although veteran players in Malaysia lack proper mainstream knowledge, they learn from the bottom and gain experience along their endeavours to reap the excellence. Future housing entrepreneurs with advanced educational backgrounds may perform better and be able to boost up their business performance under the guidance of senior and experienced mentors.

Obviously, education is important to improve the entrepreneurial competencies. Proper education will supply technical knowledge to homebuilders. Informal education can be enhanced through social networks and experiential learning within the real estate environment. Meanwhile, prior experience is also valuable to housing entrepreneurs in enabling heightened levels of business success. Either management experience or industry-specific experience, all related experience might be useful to act as practical knowledge for the housing developers. Working experience will allow an entrepreneur to judge a situation more clearly to achieve their goals in business (Gilmore, Carson, & O'Donnell, 2004), and it will implicitly minimise the risks and chances of failure. Besides that, past experience also influences on where and how the firm should respond. As one of the unique resources, experience in similar type of business will provide the firm with competitive advantage that is difficult to imitate.

1.3.1.2 Positive Traits

Personality and performance may be intricately correlated, since personality may have a direct influence on the leadership ability and style of individuals as well as team performance and the overall effectiveness of housing organisation. Inculcating excellent characteristics on the owner-managers will be helpful to sustain success into the business. Personal attributes such as hardworking, tenacity, honest, proactive and highly confident spur the development of interior spirit. For instance, through high determination, the mindset will intuit solutions to overcome upcoming problems.

It is quite fascinating to see that some housing developers tend to set annual and long-term stretch targets for each of their projects. To become visionary entrepreneurs, residential developers need to adjust specific and targeted goals on what they want to achieve from their property venture as well as step-by-step outlines of how they will accomplish their projected objectives. On top of that, housing developers should be imbued with the elements of entrepreneurial orientation such as proactiveness, innovativeness and risk-taking propensity. In short, a combination of positive psychological characteristics is deemed to affect the entrepreneurial performance of Malaysian housing developers.

1.3.2 Theme 2: Firm Resources and Management

1.3.2.1 Land Bank

In order to move forward, housing developers need to assemble suitable lands for future development. Land bank portrays firm survivability in the industry since more lands acquired depict greater strength of the company. Although land cost constitutes the major share of construction costs, the key to survival lies on land banking ambitions. To that extent, successful housing developers will try to embrace land banking strategy for business survival. However, purchasing parcels of land involves investment and locational decisions that will ascertain the project success afterwards. In Malaysia, land acquisition can be realised through own initiative or through joint venturing with landowners.

1.3.2.2 Strategy and Planning

To achieve considerable outcomes, housing firms adopt different strategies and plan their business strides wisely. In certain situations, they have to embrace a grand overarching strategy to sustain their business. It is interesting to see some housing developers tend to outsource the professional services, while others choose to sublet certain aspects of construction processes. Careful plans and tactful strategies will allow the firms to focus on their projects and reduce the overhead costs due to minimal staff recruitment. The development of business plan should be in parallel with clear time planning and goal setting, so that the opportunities of market-driven business can be easily retrieved.

It is the uniformity of the intended strategy to organisational structure, technical and managerial supports, which makes firms progress. The organisational structure of housing developers has to reflect and support the envisaged strategy. Only companies with clear leadership and vision will be successful in the long run. Strategies that are unique to housing developers such as joint venturing in developing the housing estates and outsourcing the professional services are different between large and small organisations. Pragmatically, in order to stay competitive and be socially responsible, housing developers may adopt creativity and value-adding elements in their housing projects.

1.3.2.3 Market Analysis

Understanding market and customers' preferences will be valuable to housing entrepreneurs in order to remain aggressive in the marketplace over the long period. Keeping abreast of current trends, including any changes in consumer spending habits, mortgage rates and the unemployment rate, to name a few, enables residential developers to acknowledge current conditions and subsequently plan for the future. Housing development can be a highly lucrative business but can also result in heavy losses as well. To succeed in the real estate business, housing developers must recognise potentially rewarding opportunities and predict market trends. In relation to business plan, it is essential to tailor the development with the demand in the proposed area. The successful housing developer is necessarily a shrewd salesperson with a deep knowledge of property markets and a broad understanding of the contracts, laws and tax regulations that apply to real estate transactions.

1.3.2.4 Financial Management

A good housing developer must not only be a big picture thinker, but ought to know when it is important to take care of the details especially on financial matters. Among the usual costs to be observed include the cost of the delivery, available financing during project construction and the final delivery cost to the buyers. Businesses in real estate operate on cash flow as it provides strength to both project and organisation. Works in progress require cash in hand to be transformed into project completion. Sound budgeting and costing as well as prudent cash flow management are among the fiscal accounting principles that must be handled meticulously by the housing entrepreneurs.

1.3.2.5 Social Network

Typically, housing developers work with many different counterparts along each step of development processes, including architects, planners, engineers, surveyors, estate agents, contractors and government agencies. Apart from forming relationships with the said construction professionals for the sake of timely and quality work in the future, housing developers should build relationships with local and independent lenders to finance their investment activities for the provision of swift and personalised service at competitive rates. On top of that, professional networks that are based on mutual trust and strong tie have been proven to influence business survival and growth. Good relationship with industry stakeholders will also expedite the approval, assist in fund acquisition and enhance knowledge management.

1.3.2.6 Marketing Principles

Undeniably, sagacious marketing could double demand for the product. The onus is not only on the administration to come up with a wise marketing strategy, but the members of organisation should work hand in hand to push the product sales. That is why housing developers should put smart and planful marketing into effect to deter any potential shortcomings. In every entrepreneurial activity, marketing mix is the most vital step to be considered for the business success. The most important thing is how the product is placed in the market. Yang and Zhu (2006) believe that customer's recommendation through word-of-mouth promotion is the best marketing approach and more efficient than other advertisement. At the dawn of the importance of the Internet, marketing strategy is getting complex and diverse. In practice, outstanding marketing policy is capable to build superior image and branding for future undertakings.

1.3.2.7 Quality Products

Undoubtedly, the performance of housing estate and its quality and facilities can affect the residences' quality of life and encourage satisfaction among the buyers (Colesca & Alpopi 2011). Production standard must be well defined in conformity with high specifications, tighter tolerances and fewer faults or blemishes. Quality workmanship with good after-sale services contributes to profit and market expansion. Through quality product and service, entrepreneurs may enjoy plentiful benefits in terms of good reputation, financial savings, etc. In a nutshell, the ethos of successful housing developers is to provide a high-quality house that is always going to be distinguishable in design while meeting purchasers' demands, needs and requirements.

1.3.2.8 Innovative Projects

As lifestyles evolve, purchasing patterns change and transform in accordance with current trend and modernisation. To remain nimble in business, innovation in terms of design and administration must be implemented. It would be great to note the differences between large-scale housing developers and small-scale homebuilders in terms of their innovativeness and applied innovation. McCoy, Thabet, and Badinelli (2009) admit the reluctance of homebuilders to embrace new technologies in their working style and product design due to various uncertainties. Notwithstanding the unwilling initiatives, interest in the green concept is still growing as more developers are adopting green features in their developments. While green homebuilding is a new national trend in Malaysia, having product innovation would help to steer the company through higher performance. After all, development is more than just designs; it is also the functionality, viability and capability to develop the project with attractive concept.

1.3.2.9 Project Location

Based on property value determination, houses located in good location will attract more purchasers and investors. Buyers nowadays are getting fussy and their desires are more complex to be fulfilled. A number of spatial aspects such as shopping accessibility, educational and recreational facilities as well as other neighbourhood amenities seem to be the most sought-after criteria to be accounted for the best residential area. Hence, housing developers must be able to spot an area that is on the rise and buy early, when there is the greatest chance of making a healthy profit. They should try to look for areas of growth and gentrification where other developments are taking place or are planned. At present, in terms of regions in Malaysia, Klang Valley, Johor Bahru and Penang are the most strategic locations with great demand of houses.

1.3.2.10 Firm's Reputation

The ability of housing developers to deliver and complete the houses within the stipulated period will influence and act as a benchmark for the homebuyers' appraisal. Sterling name and track record will put an increase in buyer's confidence. Unique reputation will give competitive edge upon competitors, engender buyers' loyalty and infuse company's image in their mind. Housing developers in Malaysia have recognised high corporate reputation to be linked with sustained financial performance over relatively long periods. Branding image of housing firms will arrive once the projects attract warm responses that demarcate hard works from the owners and team members. Comparatively, experienced buyers are inclined to purchase houses constructed only by reputable companies that deliver structurally sound projects on schedule, within budget and of good quality.

1.3.2.11 Housing Price

Li, Cheung, and Sun (2015) allude to the pricing strategy as a difficult decision made by the housing developers due to changes in homebuyers' preferences and affordability. Issues on underpricing and overpricing are rather pertinent to be considered by housing developers. Since the buyers nowadays are sensitive to price, housing developers must be wary and cautious in their pricing policy. Pricing determination must consider the theory of demand and supply based on thorough market study. It is relatively true to say that housing price has a positive relationship with size of house, vocational attributes, neighbourhood concept and other applied innovations. Apparently, housing developers must be prudent in pricing their products, or they may drive away their potential buyers. It should be noted that homebuyers do not look at house prices only because they might also consider other factors such as conducive environment, easy accessibility and vicinity safety.

1.3.3 Theme 3: External Factors

1.3.3.1 Government Roles

Government directly intervenes in the housing market by regulating the house prices and enforcing new housing policies. To build houses and involve in real estate development, housing developers need to be responsive with the policy changes and sensitive to the government's regulations for the sake of business growth. They have to interact with state or local authorities and make further negotiation with the government officials to expedite related approvals and other planning processes. Since housing developers encounter very lengthy process in getting approvals from the local authorities, the federal government has put an effort to improvise the system in order to ensure the procedures will run smoothly. Without doubt, satisfactory government support plays a meaningful role in business success because government possesses a big say in the approval procedure of numerous stages of the development process.

1.3.3.2 Macroeconomic Indices

Housing development sector is most vulnerable to fluctuations of the macroeconomic situation that warrant real estate companies to undertake more radical measures during the crisis. Generally, it deals with the economic indices that affect the housing market. Sometimes, the warning signs were unavoidable or overlooked by the housing developers when they had to launch or sell their products during bad times. During period of volatility and uncertainty, some housing developers prefer to defer their project launch until stable condition. Essentially, property developers must be aware with the fluctuations of economy to ensure that they were not caught up in the terrible mood of times. Responses to the recession may be varied by firm size since larger housing firms might adopt different market segmentation tactics in comparison with smaller counterparts. Most Malaysian housing developers experienced the bubble economy and were tremendously influenced by the aftermaths, becoming more cautious in their outlook. Even though the crisis might proffer source of opportunities, it normally poses a serious threat for the existence of new homebuilders. Therefore, timing of land acquisition and property sale has been paramount to developers in weathering the cyclical property market.

1.3.3.3 External Fund

Bear in mind that becoming a property developer requires a lot of money. Businesses whose owners can access funding sources are less likely to fail. It is essential for the newcomers to raise the necessary finance once they intend to delve into this industry. Banks and other financial institutions will have dedicated property finance experts that can advise them on available and suitable funds. Inspired entrepreneurs who intend to dabble in property development industry must be ready with adequate capital in order to surmount the risks and challenges. For private housing developers, the light-bulb moment must be followed with the long hard slog of strategy and marketing that encompass financial, social and human capitals. Normally, the majority of Malaysian housing developers rely on external funding to finance their projects. Personal savings may be useful to add or back up any unforeseen expenses.

1.4 Summary

An understanding of the customers' needs, proper building techniques, creative marketing efforts and solid financial management procedures will always be the deciding factors in a business success of housing developers. In particular, housing developers should build houses distinctively with high level of passion because the sense of accomplishment and satisfaction actually drives them towards greater performance. In order to succeed, they should not only set new benchmarks and stay ahead of the market, yet they have to develop a portfolio of outstanding developments that are unique, resulting in high returns on investments for their purchasers. As such, there is urgency for the researchers to find the best criteria and elements of successful housing developers to be shared with fellow industry players and nascent property entrepreneurs.

References

Abdul-Aziz, A. R., Ho, S. Y., & Jaafar, M. (2006). Competitive resources of private housing developers: The Malaysian perspective. *Journal of Engineering, Design and Technology, 4*(1), 71–80.

Barney, J. (1991). Firm resources and sustained competitive advantage. *Journal of Management, 17*(1), 99–120.

Colesca, S. E., & Alpopi, C. (2011). The quality of Bucharest's green spaces. *Theoretical and Empirical Researches in Urban Management, 6*(4), 45–59.

Gilmore, A., Carson, D., & O'Donnell, A. (2004). Small business owner-managers and their attitude to risk. *Marketing Intelligence & Planning, 22*(3), 349–360.

Li, L. H., Cheung, D., & Sun, H. (2015). Does size matter? The dynamics of housing sizes and prices in Hong Kong. *Journal of Housing and the Built Environment, 30*(1), 109–124.

McCoy, A. P., Thabet, W., & Badinelli, R. (2009). Understanding the role of developer/builders in the concurrent commercialization of product innovation. *European Journal of Innovation Management, 12*(1), 102–128.

Porter, M. E. (1985). *The competitive advantage: Creating and sustaining superior performance*. New York: Free Press.

Teece, D. J., Pisano, G., & Shuen, A. (1997). Dynamic capabilities and strategic management. *Strategic Management Journal, 18*(7), 509–533.

Yang, S., & Zhu, Y. (2006). Customer satisfaction theory applied in the housing industry: An empirical study of low-priced housing in Beijing. *Tsinghua Science and Technology, 11*(6), 667–674.

Chapter 2
Profile of Malaysian Housing Developers

Keywords Malaysian housing industry • Residential developers • Foreign and local players • Niche strategy • Merger and acquisition

2.1 Introduction

Apart from construction, housing development is regarded as the most influential industry in Malaysia. Since housing is a crucial segment of real estate, it is somehow interwoven with the construction sector in terms of its generic background and industry foundation. Due to the similarity of the two businesses, many construction firms have expanded into property development. During buoyant economic times, property development has captured the imagination of Malaysian businesses across the board especially in the last decade. Almost every sizeable company has a property division, and almost all businesses with extra cash and land bank are venturing into property development. Even landowners without adequate capital are joint venturing with developers to ensure they would get to share this seemingly expanding yet lucrative piece of cake. Great demand from the public towards the housing provision has created a room of opportunities for homebuilders to explore this challenging business.

Undoubtedly, the need for development is steadily increasing due to population growth and the changing tastes of how the populace wants to live. The property market in Malaysia has been on an upward rising trend for the past 10 years and as reported by the National House Buyers' Association, Malaysians will continue to embrace residential and commercial property purchases as they look for various ways to enhance their living conditions and investment portfolios. In Peninsular Malaysia, property development activities are concentrated in major urban areas namely Klang Valley, Putrajaya, Penang and Johor Bahru, all of which are areas of high population concentration and employment opportunities. However, since the Klang Valley is saturated with high-end developments, some analysts also foresee the northern part of Negeri Sembilan to be the next investment hub after Klang Valley.

Based on the current situation, housing projects are explicitly concentrated in the urban and suburban areas, where the purchasing power is higher and the market is extensive. According to Shuid (2015), urbanisation levels also increased significantly

M. Jaafar et al., *Business Sustainability Model for Malaysian Housing Developers*, DOI 10.1007/978-981-10-5266-8_2

from 34.2% in 1980 to 71% in 2010, and this is expected to reach 75% by the year 2020. It therefore presents a huge opportunity for potential residential developers to position themselves in meeting the expectations to suit the present standard of living and modern preferences. To further maximise the return on land and match with the preferred lifestyle of living in urban Malaysia, the strata format has gained substantial momentum especially in inner-city locations. Therefore, the need for innovation among housing developers is extremely overriding for the sake of excellent planning of townships and the realisation of high-quality residences.

2.2 Brief History of Housing Developers in Malaysia

Housing provision in Malaysia is executed by both the public and private sectors. Generally, the housing projects constructed by the public sector are normally financed by the government through funds that have been allocated under the national budget (Sirat et al., 1999) for the implementation of public housing programmes, institutional quarters and other administrative constitutions. Since the early period after independence, the private sector has played a central role in providing houses for the populace. Private sector housing consists of houses erected by private developers, cooperative societies and individuals or a group of individuals. Their housing schemes are normally self-financed, deposited through prospective buyers or obtained from various financial institutions (Sirat et al., 1999). The participation of private sector was augmented in the 2nd Malaysia Plan, started circa 1970, when the government sought the cooperation of private developers to provide low-cost houses by making it mandatory for them to build at least 30% low-cost houses in every housing project.

According to Jaafar and Sahari (2013), ethnic Chinese is the principal player in the Malaysian housing industry. Other races like the Malays and Indians account for a small portion of the sector with minimal presence in most parts of Malaysia. The involvement of Malays in the real estate industry, for example, was officially embarked after the implementation of the New Economic Policy in the 1970s. Often, as private housing developers, they must obtain a licence from the Ministry of Urban Wellbeing, Housing and Local Government (MUWHLG) and are encouraged to register as members with the Real Estate and Housing Developers' Association (REHDA) Malaysia (Jaafar & Ali, 2011). REHDA is the representative body for property developers in Peninsular Malaysia, whereas Sabah Housing and Real Estate Developers' Association (SHAREDA) corresponds to Sabahans, while Sarawak Housing and Real Estate Developers' Association (SHEDA) signifies the Sarawak chapter. The associations are recognised as an official channel of communication with other stakeholders on matters relating to the building and construction industry.

2.3 Housing Development Process

In the Malaysian housing industry, private sector participation is administered by both federal and state governments. They are responsible for formulating related policies, laws and regulations with respect to housing construction (Shuid, 2015). Basically, the housing development process involves three main stages with the most crucial part being the planning phase. A developer has to firstly obtain all the planning approvals before any physical work can commence on site and prior to the issuance of any advertising permit by the relevant authorities. After the completion of pre-development process, the construction stage will begin, followed by the post-construction stage. In the meantime, while waiting for land conversion and site approval, developers may need to source for bridging loans from financial institutions and seek credible lawyers to prepare the legal documents which include the S&P (sale and purchase) agreement to be signed by the purchasers once the project is launched to the market.

One of the respondents from Jaafar and Sahari (2013) recommended nascent entrepreneurs and new housing players to promptly adopt industry knowledge and be aware of the bureaucracy to avoid any future fraud against their projects. To carry out their housing development, residential developers need to alter land use for development purposes, integrate resources required for the projects, interpret the demand and supply requirements of the market, provide the housing units and finally sell the completed houses. Invariably, potential newcomers who are looking forward to get involved in housing development in Malaysia must follow all the regulations imposed by the government, starting from obtaining approval for planning permission, and other construction rules that govern the activities from the inception to completion stages before obtaining the Certificate of Completion and Compliance (CCC). Apart from that, with the current "sell-then-build" practice, housing developers normally have to ensure that a project has met certain predefined milestones before any housing units can be put onto the market and sold.

2.4 New Big Players and Mergers in the Malaysian Housing Industry

Despite stiff competition experienced by the existing industry players, new entrants abound in the housing landscape every year, encompassing small-scale entrepreneurs and large business players. While new housing developers are usually small in size and lack experience, others who have vast prior networks prefer to embrace strategic approaches through merging and joint venturing. Besides that, as the availability of development land is more constrained in Malaysia, one of the most effective strategies for securing land is through the takeover of other firms to immediately access their land banks, leading to further consolidation. Essentially, the significant rationale for a merger is to create a bigger land bank, with the added advantage that

Table 2.1 Main business of non-property-related companies

Organisation's name	Original core business
Anzo Holdings Bhd	Timber door maker
Caely Holdings Bhd	Lingerie retailer
Fiamma Holdings Bhd	Electrical home appliances distributor
GSB Group Bhd	Optical disc manufacturer
Luster Industries Bhd	Plastic injection moulding manufacturer
Milux Corporation Bhd	Household appliances distributor
PRG Holdings Bhd	Furniture components manufacturer
Scanwolf Corporation Bhd	Plastic extrusions designer and manufacturer
Spring Gallery Bhd	Ceramics maker
Weida (M) Bhd	Building materials maker

the enlarged entity combines the different experiences and product offerings of the two companies.

In Malaysia, the most prominent merger occurred in 2010 when the government-linked company, UEM Land Holdings Berhad decided to integrate their businesses with a leading high-rise builder, Sunrise Berhad with a newly-renamed entity, UEM Sunrise Bhd. More recently, in 2015, OSK Holdings Berhad completed the corporate exercise to merge the businesses and operations of PJ Development Holdings Berhad and OSK Property Holdings Berhad into the OSK Group. It is deemed that the merged entity is able to be better placed when competing against the bigger peers and foreign developers which are abundantly coming into Malaysia. While smaller players might continue offering niche projects, the bigger players would have better financial strength to focus and build even larger scale developments in new and untapped areas.

Apart from external acquisitions, a new scenario of diversification is occurring in Malaysia. Although most investors would prefer a company to concentrate only on their core business, umpteen non-property related companies on the Bursa Malaysia (Malaysian Bourse) have diversified into the lucrative property development as part of plans to expand their income stream. For these huge organisations, the real estate industry offers new dimensions to explore opportunities beyond their core business activities. Public-listed companies expect long-term profitable outcomes through diversifying their earnings base and delivering greater value to the shareholders. Others see this opportunity as massive economic prospects in the land development business especially during the "boom" period.

The following table portrays the public-listed companies that have recently delved into property development (Table 2.1).

For illustration, Luster Industries Bhd, through its unit, Pembinaan LSP Jaya Sdn Bhd, has formed a joint venture (JV) with Koperasi Hartanah Malaysia Bhd (Kohama) to embark on a housing project in Pengkalan Hulu, Perak. Scanwolf

Corporation Berhad, on the other hand, has carried out a mixed development project in Kampar, Perak, which comprises commercial suites, townhouses, two apartment blocks and a hotel, to be undertaken by its indirect unit, Scanwolf Development Sdn Bhd. Another instance is Fiamma Holdings Bhd that has successfully built "East Parc@Menjalara" in Bandar Menjalara and "Vida Heights" in Johor Bahru. Also, in 2013, a foreign-based firm, AMS Light Metal Sdn Bhd (wholly owned by Asia Metal Solutions Pte Ltd. from Singapore), ventured into the property development business via the takeover of Goldrich Concept Sdn Bhd, a licenced Malaysian developer, and launched its maiden project, i.e. the first luxury gated and guarded landed property development in Pekan Nanas, Pontian.

Meanwhile, a renowned Malaysian textile retailer, Jakel Trading Sdn Bhd, recently launched its maiden residential property project in Cheras, Kuala Lumpur, in April 2015 called J.Dupion which was inspired by the name of an extraordinary and expensive silk, a material which was famous in Europe during the times of the Renaissance. Jakel had been involved in the real estate arena since 2003, pioneered by the late Founder, Mohamed Jakel Ahmad, when they started with buying shop houses for leasing and bought plots of land as investment. Since 2009, during the economic downturn, they started to accumulate land more aggressively. Jakel decided to venture into property development because of the tremendous opportunities, especially since its property investment of more than 10 years has been contributing significant recurring income.

On a separate note, it is a common situation in Malaysia to see building contracting firms dabbling in property development. Most firms started their business as a small contractor, and the move into property development is seen as synergistic to its construction and engineering businesses in order to enlarge business operations and earnings base. In Ho Hup Construction Company Bhd's example, the construction firm foresees property development to be its main driver going forward and has mapped out plans to develop the 60-acre Bukit Jalil land in Kuala Lumpur, which will have a gross development value of RM1.05 billion. Ho Hup is developing 10 acres of this site on its own and the remaining 50 acres will be jointly developed with Malton Bhd, a major property player.

Above all, venturing into other industries must be judiciously planned in order to materialise the expected profit. At the end of the day, housing entrepreneurs should look at what is scalable and available in the market. Most importantly, they need to have relevant expertise and sufficient land bank to sustain the business. Four aspects must be meticulously observed, i.e. encompassing the right person to manage the right product, at the right location with the right price. For the record, among the property players who have excelled in multiple industries are IOI Corporation Bhd, Mah Sing Group Berhad, KSL Holdings Bhd, Scientex Berhad, South Malaysia Industries Bhd, Knusford Berhad, YTL Corporation Bhd, DKLS Industries (Malaysia) Berhad, GUH Holdings Bhd, JMR Conglomeration Bhd, Hock Lee Holding Sdn Bhd, Hock Seng Lee Berhad and Shin Yang Corporation Berhad.

2.5 Perceived Success of Malaysian Housing Developers

As business failure in housing development industry is normally associated with issues on property overhang and oversupply as well as abandoned housing projects, the mechanism to distinguish business success is seemingly unknown and hard to be justified. Perhaps the effort taken by The Edge Communications Group to create awards and list the top property developers might fill some of the gap and address part of the issue. The concept of success is rather similar to the idea of striving for the number one position in the market, a sort of industry "Oscars". By referring to the corporate website (www.theedgemalaysia.com), it revealed that the anchor awards of The Edge Property Excellence Awards was established in 2003 to rank Malaysia's best property players from the consumer's perspective based on their quantitative and qualitative attributes. For reference, the "qualitative" attributes include the quality of the product, innovation and creativity, value creation for buyers, image and expertise, while the "quantitative" elements were the companies' shareholders' funds, revenue, pre-tax profit and net gearing for the fiscal year ended. More recently, two additional criteria were included comprising financial strength of the developers and consistent delivery of quality products to customers. All companies listed in the property sector of Bursa Malaysia are automatically considered for the awards without having to make a submission, while other non-listed development firms need to prepare some organisational documents and records for evaluation. From simple observation, three leading listed companies, i.e. S P Setia Group Berhad, Sime Darby Property Berhad and Sunway Berhad, appear to be the most dominant property developers who have consistently outperformed other contenders (based on the yearly ranking) since the beginning of the conferment.

Award recognitions aside, there is another measurement to gauge the success of property developers. As a symbol of success, big conglomerates always look forward to be publicly known. A few local housing entrepreneurs have set their target to be listed on the Main Board of Bursa Malaysia Securities. To get listed on the Bursa Malaysia (previously known as the Kuala Lumpur Stock Exchange), a company needs to meet certain requirements. For housing developers, they are required to possess a minimum land bank of 1000 acres, have sufficient ongoing property development projects and obtain an aggregate after-tax profit of not less than RM30 million for the 5 full years prior to listing. Large companies often become public limited companies (PLCs) to gain better access to capital from the financial institutions. Therefore, to be listed in the local stock market in Malaysia, housing developers must have a large portfolio of land bank and have a consistently profitable track record.

2.6 Current Scenario of Housing Developers in Malaysia

Before dwelling on further, few terminologies such as "boutique developer" and "niche market" must be defined to provide clearer picture and offer better understanding towards ensuing descriptions. A boutique developer is typically small in size and offers highly specialised services or products. The houses are delivered in a unique atmosphere and style that caters to the particular tastes and preferences of the clientele. Boutique developers generally do not own large land banks; once they acquire a plot, they will work hard to achieve a fast turnaround. On a different note, a niche market deals with a specialised market which is relatively small. This type of marketing effort entails a particular group of people through market concentration and well-defined segment of population.

By adopting niche marketing, housing developers usually serve the specific and unique needs of narrowly defined buyers. Interestingly, boutique property developers normally focus on niche markets, with an eye on medium- to high-end developments in prime locations. In the Klang Valley especially, the existence of boutique developers is very significant as they are only keen to develop smaller plots of high-quality projects, spanning from high-end serviced apartments to affordable homes. In short, specificity and focus are their main business objectives.

The subsequent discourse reveals the current situation of Malaysian housing developers according to the various states of the country. Since more than a thousand housing developers exist in Malaysia, not all profiles of developers are mentioned as only active players will be documented. The authors may have overlooked certain areas and homebuilders due to the lack of exhaustive list and inadequate sources.

2.6.1 Johor

Johor is very fortunate to have Iskandar Malaysia in its territory. On top of that, the state's economy also enjoys its proximity to developed nation, Singapore, as Singaporeans tend to have shopping in its capital, Johor Bahru, due to low exchange currency. When the Iskandar Malaysia growth corridor was initiated, development projects within the region enjoy substantial public demand. The construction of new towns increased the production of formal housings by private developers. Apart from Nusajaya that undergoes exponential development, other hot spots such as Pasir Gudang, Tebrau, Kempas, Skudai, Tampoi, Kulai and Senai also experience impressive expansion. Since 2011, strata properties are mushrooming around the city of Johor Bahru. At the moment, the primary market for high-rise residences can be captured in Skudai, Tampoi and Kempas.

Among notable property developers that have invested in Iskandar Malaysia are UEM Sunrise Bhd, Sunway Bhd, Mulpha International Bhd, WCT Holdings Bhd and United Malayan Land Bhd. Other conglomerates (with project names) such as Eco World Development Group Bhd (Eco Spring, Eco Botanic and Eco Summer), IOI Properties Group Bhd (Bandar Putra Kulai and Taman Kempas Utama) and Scientex Berhad (Taman Scientex in Pasir Gudang and Taman Mutiara Mas) also exist with big pockets of funding capital. Not to be left out, government-linked companies such as Johor Land Berhad and UDA Land (South) Sdn Bhd also own few projects in this vicinity.

For a record, at the moment, two of the largest developments in Iskandar Malaysia are the 1800-acre Sunway Iskandar by Sunway Bhd and 4500-acre Gerbang Nusajaya by UEM Sunrise Bhd. Local players such as Daiman Group, MB Group (formerly known as Mahabuilders Sdn Bhd) – a construction company-turned-property developer, Keck Seng (M) Bhd, Lian Ak Group and Binaik Equity Bhd are among the major players in Johor. KSL Holdings Bhd, in particular, besides its strong presence in Iskandar region, has ventured to Klang Valley as well. Other local property tycoons such as Gromutual Berhad (with projects in Melaka) and BCB Berhad (Kluang-renowned homebuilder who expands to Klang Valley) as well as Joland Group are looking to enlarge their business to other states specifically in Greater Kuala Lumpur. Apart from that, medium-sized housing developers who specialise in the development of high-quality residential property development such as Erawish Development Sdn Bhd, Multimax Development Sdn Bhd, Berinda Properties Sdn Bhd, Austin Heights Sdn Bhd, Rawhide Sdn Bhd and Scudai Development Sdn Bhd also make great impacts towards the advancement of Johor.

For illustration, by generating vibrant housing and commercial communities, Lian Ak Group has generated a wide range of experience and expertise. Being well-versed in the field, Lian Ak Group has transformed from an impressive standing as the leading developer in Southern Malaysia, sprawling over Johor Bahru, Segamat, Kluang, Tangkak, Yong Peng, Endau and Melaka. Noticeably, there are niche developers in Johor Bahru as well, such as Alpha Astral Realty Sdn Bhd, Front Concept Sdn Bhd and Northstar Frontier Sdn Bhd that focus on certain market segments and groups.

Outside Johor Bahru, a number of small and medium developers can be traced such as Weeco Kemajuan Sdn Bhd (in Yong Peng) and Sia Her Yam Realty Sdn Bhd (in Segamat). On top of that, Muar, Batu Pahat and Kluang have become popular destinations for many tourists especially for those from Singapore, in transit to other parts of Peninsular Malaysia. Ideahome Development Sdn Bhd, KCC Development (M) Sdn Bhd, Sunbay Realty Sdn Bhd, Gold Li Development Sdn Bhd, Lishung Holdings Sdn Bhd, Sawit Warisan Sdn Bhd and Lancar Harta Sdn Bhd are among the famous housing developers in Muar, while Mudi Pembinaan Sdn Bhd, Orchard Vest Sdn Bhd, Jayafuda Development Sdn Bhd, JBBP Development Sdn Bhd, Gandingan Satu Sdn Bhd, Jinyuan Property Sdn Bhd, QBS Development Sdn Bhd and Countryland Realty Sdn Bhd have become the household names for populace in Batu Pahat. Somehow, among the leading housing developers in Kluang include Majupadu Development Sdn Bhd, Everwin Realty Development Sdn Bhd, Kluang Setia Development Sdn Bhd and Kemajuan Tong Tor Sdn Bhd.

Although Hong Leong IB Research reported that Johor region remains oversupplied in near term with incoming supply and planned supply amounting to 47% of existing stock, many outside business tycoons still believe on the credibility of Iskandar Malaysia as they are not reluctant to invest in Johor. Interestingly, foreign developers such as Guangzhou R&F Properties Co Ltd. (from Hong Kong) and Hao

Yuan Investment Pte Ltd. (from Singapore) have bought some parcels along the straits in Danga Bay to develop high-end buildings.

2.6.2 *Melaka*

Housing development in this historical state is dominated by local players such as NKS Development Sdn Bhd, Teobros Development Sdn Bhd, GJH Construction Sdn Bhd, Unitedhome Development Sdn Bhd, PB Realty Sdn Bhd, Grandhome Development Sdn Bhd, Teladan Setia Sdn Bhd, Faithview Development Sdn Bhd, Bukit Kenyalang Development Sdn Bhd, Cheng Xing Development & Construction Sdn Bhd, Chip Hock Realty & Development Sdn Bhd, JS Group, Sri Pengkalan Binaan Sdn Bhd, Pleasant-Ville Development Sdn Bhd and Jaymuda Group due to several quirky rules imposed by the state itself like issues on Melaka Customary Land (MCL) and other ownership procedures. Principally, these local developers are operating in and originally from Melaka.

Bigger firms such as Hatten Group and Asia Selatan Group are also Melaka-based real estate developers. Asia Selatan Group, for instance, is a reputable and reliable property developer in the state of Melaka with vast experiences in timely deliveries of quality industrial, commercial and residential projects since 1980. Their housing projects are located not only in Melaka (comprising projects in Bemban, Umbai, Jasin, Merlimau and Kesang) but in Gemencheh, Negeri Sembilan, as well. Notwithstanding the locality issues that hinder the interest of non-local homebuilders, few large companies are able to present in this state such as MTD Capital Bhd through Metacorp Properties Sdn Bhd for Taman Tasik Utama and Scientex Berhad through Scientex Heights Sdn Bhd for Taman Muzaffar Heights.

2.6.3 *Negeri Sembilan*

Despite a number of property developers from Klang Valley venturing into this state, there are still quite a handful local residential developers exist. Among the active housing developers in Negeri Sembilan include Assurplus Sdn Bhd, Berjasa Century Construction Management (M) Sdn Bhd, Chin Ho Realty Development (M) Sdn Bhd, Heap Hoe Lee Holdings (M) Sdn Bhd, Mega 3 Housing Sdn Bhd, Seri Pajam Development Sdn Bhd, Dataran Segar Sdn Bhd, Prestajaya Sdn Bhd, Primuncak Properties Sdn Bhd, Sunrise MCL Land Sdn Bhd and North Elegance Development Sdn Bhd. Mega 3 Housing Sdn Bhd, for example, have five branches situated in Seremban, Negeri Sembilan; Masjid Tanah, Melaka; Triang, Pahang; Sepang, Selangor; and Telok Panglima Garang, Selangor.

Besides that, large companies such as Nilai Resources Group Sdn Bhd and Matrix Concepts Holdings Bhd are also originally from Negeri Sembilan. Through its subsidiary, BBN Development Sdn Bhd has developed vibrant township called Putra Nilai, while Matrix Concepts Holdings Bhd is known for its township projects, namely, Bandar Sri Sendayan in Seremban and Taman Seri Impian in Kluang, Johor. Several other significant developers such as Gunung Nuang Group, Bonus Timur Sdn Bhd and Gaya Idaman Sdn Bhd are focusing in rural areas, specifically in Bahau, Gemencheh, Tampin and Gemas.

Also, due to its strategic location as being so close to Kuala Lumpur, large property conglomerates from Kuala Lumpur have ventured in Negeri Sembilan as well. For instance, IJM Land Berhad through Seremban Two Holdings Sdn Bhd and Sime Darby Property Bhd through Sime Darby Ainsdale Development Sdn Bhd with projects located in the northern part of Negeri Sembilan. Interestingly, medium-sized housing developers from Kuala Lumpur also own few projects in Negeri Sembilan such as Ehsan Plant & Property Sdn Bhd who have marked its stamp in Tampin.

2.6.4 *Klang Valley (Comprising Selangor, Kuala Lumpur and Putrajaya)*

Housing market in Klang Valley is very competitive owing to the presence of big players with massive land bank. Nonetheless, for small and medium homebuilders, there are not many opportunities in the Klang Valley due to scarcity of land. So, most of their projects tend to be situated outside the vicinity. Kuala Lumpur and Selangor remain the national core market with more balanced demand and supply dynamics as population from the centre of Greater Kuala Lumpur moving up north to places like Rawang which was formerly a quiet town and towards the south-east to as far as Semenyih.

A handful of niche property developers exist in Kuala Lumpur, focusing on developments of high-end integrated commercial and boutique residential projects. They are, among others, Amber Homes Sdn Bhd, Pinnacle Homes Sdn Bhd, Bellworth Developments Sdn Bhd and J&C Homes Holdings Sdn Bhd. Meanwhile, in Selangor, Gabungan AQRS Berhad, a niche lifestyle property developer that was listed on Bursa Malaysia on 31 July 2012 also becomes a major homebuilder in Klang Valley with its variety of products. Besides that, boutique development companies such as Amphil Corporation Sdn Bhd, Dawntree Properties Sdn Bhd, Kueen Lai Group, Juta Asia Corporation Sdn Bhd, Simas-D Sdn Bhd, SYF Development Sdn Bhd, DA Land Sdn Bhd and OCR Land Holdings Sdn Bhd also play a remarkable role in Klang Valley.

Another fascinating segment is pertaining the existence of Bumiputera housing developers especially in Selangor. Bulatan Mekar Sdn Bhd, Darma Hemat Sdn Bhd, Perkasa Tulen Sdn Bhd, Rumpun Ehsan Sdn Bhd, Ehsan Plant & Property Sdn Bhd, Kompobina Holding Sdn Bhd, Lebar Daun Development Sdn Bhd, MDZ Development Sdn Bhd, Fairview Development Sdn Bhd, Johawaki Development

Sdn Bhd, Micro Mart Sdn Bhd, Majulia Sdn Bhd and Raya Baiduri Sdn Bhd are among the notable Bumiputera players. Ehsan Plant & Property Sdn Bhd, for illustration, is the rising residential developers that has embarked several projects in Selangor (in Sepang), Negeri Sembilan (in Tampin) and Johor (in Kluang). Its impressive performance in delivering quality properties since its inception in 2005 has marked sterling reputation towards its brand.

Since the land in the city centre is getting scarce, property developers are moving their operations to the periphery of the conurbation. Housing developers who are focusing in suburban or outskirt of Klang Valley are Amalan Setar (M) Sdn Bhd, Cheras Hong Soon Development Sdn Bhd, Estinia Group, Homecity Realty Sdn Bhd, Kamso Sdn Bhd, Meswara Group, Sin Hee Yang Property Management Sdn Bhd, Temasya Mentari Development Sdn Bhd, Trans Loyal Development Sdn Bhd, Upaya Park Development Sdn Bhd, Upicon Group, Zacklim Properties (M) Sdn Bhd, Soon Sin Hin Development Sdn Bhd and many more. Indeed, their projects have spruced up the fringes of the city centre. Other medium-sized developers which are considered fast-rising companies include Leadmont Development Sdn Bhd, KM Land Group, Binastra Group, Ehsan Bina Sdn Bhd, Kota Menara Ufuk Sdn Bhd, Iwajib Group, TCS Development Holdings Sdn Bhd, Perfect Eagle Development Sdn Bhd, Choon Tian Sdn Bhd, Yuwang Development Sdn Bhd, OCR Land Holdings Sdn Bhd and Marvelane Sdn Bhd.

Klang Valley also is the homes for the survivor of economic crisis, being regarded as long-established housing developers in Malaysia, to name a few, Sritama Sdn Bhd, Cheong Hin Holdings Sdn Bhd, Dergahayu Sdn Bhd, KLS Group, Khoo Swe Sem Sdn Bhd, Keringat Group, Lee Yen Kee (M) Sdn Bhd, Mepro Holdings Berhad, Mentari Group and LCK Holdings Sdn Bhd. Take a case of Cheong Hin Holdings Sdn Bhd, which has diversified into property development in 1980 with its maiden residential project, Taman Maju Jaya. With unblemished track record for delivering quality homes and buildings on time, they presently have projects in Serdang, Petaling Jaya and Cheras. Another example is Dergahayu Sdn Bhd, an old-time player with housing projects in all over Selangor such as Subang, Ampang, Gombak, Selayang and Cheras. With its origin and growth in Selangor, LCK Holdings Sdn Bhd also owns several small residential projects in other states such as Johor, Perak, Negeri Sembilan, Melaka and Pahang. Currently, LCK is planning to diversify its business in Kelantan by developing commercial properties in Kuala Krai. Among its new projects include Taman Kuala Selangor Utama, Taman Behrang Bestari and Taman Sri Tangkak.

Other notable huge property developers in Klang Valley are including Symphony Life Berhad, Sunsuria Berhad, Malton Berhad, TAHPS Berhad, Bandar Utama Development Sdn Bhd, Masteron Group, See Hoy Chan Sdn Berhad, A&M Realty Berhad, Selangor Dredging Berhad, Mitrajaya Holdings Berhad, Tanming Berhad and Titijaya Land Berhad. Apart from the preceding list, other prominent powerhouses in Malaysia's real estate are mostly based in Klang Valley. They are S P Setia Bhd, Sunway Bhd, Tropicana Corporation Bhd, Eco World Development Group

Bhd, Sime Darby Property Berhad, Mah Sing Group Bhd, Gamuda Berhad, Hua Yang Bhd, The Lion Group, Glomac Berhad, I&P Group Sdn Bhd, Bandar Raya Developments Berhad, GuocoLand (M) Berhad, MK Land Holdings Berhad, Sentoria Group Berhad, MCT Bhd and Miramas Group. Their excellent sales achievement, superior earnings growth prospects and strong branding have made them reliable developers of quality properties in Malaysia.

For illustration, among Tropicana's significant projects include Tropicana Metropark, Tropicana Aman, Tropicana Heights and its hugely successful 625-acre Tropicana Golf & Country Resort. It was one of the first developers in Malaysia to come up with the gated community and resort-living concept. In the meantime, Sime Darby Property Berhad is currently synonymous with some well-planned townships in Negeri Sembilan such as Bandar Ainsdale, Chemara Hills and Planters' Haven together with The City of Elmina in Selangor and Bandar Universiti Pagoh in Johor. Besides Klang Valley, Sunway owns numerous projects in Perak, Penang and Johor. In Penang, for example, their products include Sunway Cassia, Sunway Tunas (JV with Koperasi Tunas Muda Sungai Ara Berhad) and Sunway Wellesley. The most recent outstanding property developer is Eco World Development Group Bhd, with projects in three key economic regions in Malaysia, namely, Klang Valley, Penang and Johor. Among its flagship projects are Eco Majestic in Semenyih and Eco Spring in Johor Bahru. Other leading developers also have projects in various states such as Mah Sing Group Bhd (with projects in Penang and Johor), Hua Yang Bhd (owns few projects in Johor, Perak and Negeri Sembilan) and Miramas Group (completed several housing projects in Pahang, Johor, Negeri Sembilan and Selangor).

Although the involvement of private housing developers are very significant in Klang Valley, government through state-owned enterprises (e.g. Perbadanan Kemajuan Negeri Selangor) and government-linked agencies (e.g. UDA Holdings Berhad) also play a role to provide homes for the wellbeing of its citizens. They usually build affordable houses to cater the needs of the lower-income group. Aside from federal and state parastatals, some cooperative enterprises set up by certain parties have also get involved in housing development despite their trivial presence and stake.

2.6.5 Perak

From a well-known tin mining state, Perak has since been transformed into a variety of industry hubs by retaining its old charm of a food heaven and tourist destination. Its capital city, Ipoh is regarded as a retirement paradise since the costs of living and housing in Ipoh are relatively cheaper compared to other major cities in Malaysia. Plus, some analysts postulate that Ipoh is the perfect city for retirees and young married couples due to its strategic location, amenities and accessibilities. A number of medium-sized housing developers can be traced in Ipoh, Perak, such as Associated Builders & Contractors Sdn Bhd, Belkron Group, Chye Hin Development Sdn Bhd, Indra Pusaka Sdn Bhd, Kinta Saujana Sdn Bhd, Lunar Properties Sdn Bhd, Miclebina Properties Sdn Bhd, Namcom Development Sdn Bhd, Para Development Sdn Bhd, PK Lee Sdn Bhd, Pyhomes Realty Sdn Bhd, Sam Alco Land Sdn Bhd, SBG Land Sdn Bhd, KPM Corporation Sdn Bhd, Team Keris Berhad and Total Investment Sdn Bhd. They are reputable housing developers with good track records and have becoming the household names in Ipoh.

Not only in Ipoh, some of them also own projects in other districts. For example, apart from Ipoh, Villaworld Sdn Bhd also has housing projects in Bidor, Kampar and Kuala Kangsar. This trend is emulated by other fellow players such as Wawasan Pasti Sdn Bhd who owns projects in Pusing, Tronoh, Tanjung Rambutan and Taiping. Another well-experienced housing developer is Kanglian Development Sdn Bhd that has become a household name in Tapah, Bidor, Langkap and Teluk Intan. Worth to be mentioned, even though Teh & Teh Development Sdn Bhd is originally from Taiping, it also has projects in Tanjung Malim and Gerik. Similar case with T&K Property Sdn Bhd who owns projects in Teluk Intan. Comparatively, Asiabina Holdings Sdn Bhd also develops projects in Pantai Remis, Taiping and Kamunting. Another one, Perdu Pelangi Sdn Bhd, the most renowned housing developer in Kerian, a northern district of Perak, has built outstanding quality of houses and projects in Kedah, Penang and Perak.

In other district called Sitiawan, several property firms such as 586 Group, Lingbena Development Sdn Bhd, Mylander Development Sdn Bhd, Great Home Development Sdn Bhd, Setia Awan Holdings Sdn Bhd, Blue Print Properties Sdn Bhd and DKLS Development Sdn Bhd are among the prominent housing developers. From Teluk Intan, Aun Home Development Sdn Bhd, Raz Intan Development Sdn Bhd, Quantum Springs Sdn Bhd are charting their names in housing development industry. A property developer from Penang called Sunny Homes also has some projects in Teluk Intan such as Taman Desa, Taman Panglima and Taman Cahaya. Besides that, the presence of property developers from Klang Valley such as Malaysian Resources Corporation Berhad, Sunway Berhad, MK Land Holdings Berhad, Andaman Property Management Sdn Bhd, Agacia Land Sdn Bhd, Fortuneview Land Sdn Bhd and MBM Land Sdn Bhd has enliven the entire economic activities in Perak.

Locations that are near to education hubs such as Seri Iskandar, Tronoh, Tanjung Malim and Kampar enjoy significant growth of land values. Similar scenario happens to parcels or townships that are closer to industrial plots. The circumstances have created more opportunities for new entrepreneurs to jump into housing realm. New residential developers in Perak include Ngan Yin Property Sdn Bhd, Luxe Properties Sdn Bhd and Akijaya Development Sdn Bhd. Nonetheless, the state investment arm, Perbadanan Kemajuan Negeri Perak also involves in building affordable housing schemes in this state. The most recent developments of this parastatal are D'Aman Residensi and Casa Kayangan in Meru Raya, Ipoh.

2.6.6 Penang

Penang with its golden beaches, rich cultural heritage, a multi-ethnic population and a thriving commercial landscape continues to present multiple opportunities for growth. Its capital, George Town, has been listed as a UNESCO World Cultural Heritage Site in 2008 due to its unique history and heritage. Land scarcity and sky-rocketing property prices on Penang Island have driven property developers and investors towards the mainland in recent years. Based on periodical observation, Penang-grown entrepreneurial property development companies include Airmas Group, Asia Green Group, Chong Company Sdn Berhad, Ivory Properties Bhd, Hunza Properties Bhd, Tambun Land, VST Group, Oriental Max Group and Farlim Group (Malaysia) Berhad. Other Penang-based property players are Aroma Development Sdn Bhd, Aspen Group, Boon Siew Group, Lip Sin Company Sdn Bhd, PLB Engineering Bhd, Piala Mutiara Group, Cosmopolitan Homes Sdn Bhd, Emerald Capital Group, Metrio Development Sdn Bhd, Lone Pine Group, Belleview Group, Malvest Group, Syarikat Nyaman Sdn Bhd, Ujung Pelangi Sdn Bhd, Ideal Property Group and M Summit Group.

VST Group, for instance, is a family-based property development company. Besides Penang, it has expanded to other states such as Pahang, Terengganu and Klang Valley. Farlim Group (Malaysia) Berhad, on the other hand, is a prominent property developer with its flagship township known as Bandar Baru Ayer Itam on the island of Penang. This group also have residential projects in Melaka and Selangor as well as commercial development in Penang and Kuala Lumpur. Established since 1982, OHM Group is a local player with primary development on the Penang Island. It shares similar case with another veteran homebuilder, Chong Company Sdn Berhad that has built homes for Penangites since 1963. Led by Dato' Sri Dr. Lee Kuang Shing, Welcome Development Sdn Bhd is another local player who owns projects in both island and mainland of Penang.

Among formidable names in mainland are Asas Dunia Berhad, Taman Jadi Group, Bertam Properties Sdn Bhd, Loyal Greenworld Sdn Bhd, Epic Valley Holdings Sdn Bhd, Kar City Development Sdn Bhd, Metro Jelata Group, Metrio Development Sdn Bhd and GTM Land & Property Sdn Bhd. With their main office in Penang, Epic Valley Holdings Sdn Bhd is a leading housing developer in the northern region of Peninsular Malaysia with projects in Kedah, Penang and Perak. Besides that, they have also created development portfolio in Perlis, Selangor, Terengganu and Pahang as well. Owing to great demands from big population, Penang has also attracted outsiders to try their luck in property development. Large conglomerates from Klang Valley such as Eco World Development Group Bhd, Mah Sing Group Bhd, Sunway Berhad, IJM Land Berhad and S P Setia Bhd also have few projects in Penang.

State-owned enterprise, Penang Development Corporation (PDC), also plays its role in real estate industry along with other federal controlled property entities like JKP Sdn Bhd and UDA Land (North) Sdn Bhd. Notwithstanding the background of Penang is getting denser with less development land is available, new homebuilders still able to survive. Apparently, small housing developers such as Amar Iman Development Sdn Bhd, CSY Development Sdn Bhd, Northern Skyline Sdn Bhd and Tat Boot Development Sdn Bhd are healthily competing with other players to secure some parcels in Penang.

2.6.7 Kedah and Perlis

Perlis, as the smallest and least populated state in Malaysia, enjoy several shared facilities with its neighbour, Kedah. Very few state-born housing developers can be detected in Perlis that include Nepta Development Sdn Bhd, Vista Mulia Development Sdn Bhd and E-Shah Development Sdn Bhd. Since the market in Perlis is considerably small, most of them are now expanding their development projects into Kedah too. In the meantime, property developers from Kedah have also ventured to Perlis since decades ago. Among residential developers from Kedah that have projects in Perlis include AHS Realty (M) Sdn Bhd, Aima Development Sdn Bhd, Amzed Development Sdn Bhd, Encomas Sdn Bhd, LPC Land Sdn Bhd and Jubrincon Bina Sdn Bhd.

Meanwhile, for the list of active housing developers in Kedah; Aman Setia Group, Teguh Harian Sdn Bhd, Jesin Group, EUPE Corporation Bhd, Emico Development Sdn Bhd, Nice Saga Development Sdn Bhd, Bandar Saujana Sdn Bhd, Cerah Kurnia Corporation Sdn Bhd, Global Reform Sdn Bhd, GPC Development Sdn Bhd, Lean Keng Sdn Bhd, Merger Trend Sdn Bhd, Sungguh Kemajuan Sdn Bhd, Sri Awona Sdn Bhd, Rainbow Plus Sdn Bhd, Benway Development Sdn Bhd, Profit Land (M) Sdn Bhd and Percaya Jaya Group must be included. Also, quite a number of Malay housing developers can be discovered in Kedah such as Aima Development Sdn Bhd, Syarikat Sakan Sdn Bhd, PZS Development Sdn Bhd and Cash Landmark Sdn Bhd. Apart from that, state-owned enterprise, Binaraya Darulaman Berhad through its property arm, Darulaman Realty Sdn Bhd also plays a dominant role by developing township in several budding spots like Jitra and Kuala Ketil. Furthermore, large property developers from Klang Valley such as OSK Property Holdings Berhad and Plenitude Berhad have also invigorated the business scene in Kedah with some notable projects in Sungai Petani.

In actual fact, the growth of private housing developers is quite extensive in Sungai Petani, due to its close location to Penang that offers a pocket of opportunities for businesspersons to thrive. To illustrate, a housing developer from Sungai Petani, Bandar Saujana Sdn Bhd, which was incorporated in 1994, is focusing its projects in Bujang Valley only. Utilising the phase-by-phase strategy in completing its projects, the flagship projects include, Taman Lembah Bujang (Phases 1, 2 and 3), Taman Lembah Bujang Indah (Phase 4) and Taman Lembah Bujang Utama (Phase 5) which were easily sold out. On the other hand, while most developers pursue large townships in major cities, Irama Jaya Group prides itself in its property projects in quaint towns. As a form of "nation building", the Group believes that its approach to rural development will form the backbone of developing a progressive nation. Towards this end, its completed projects can be found in Sik, Merbok, Baling, Kuala Kedah and Changlun.

2.6.8 Kelantan

As a Malay-dominated state that practises Islamic teachings in daily lives, Bumiputera housing developers that are normally small in size preponderate over their counterparts. Veteran players in Kelantan include SBJ Binajaya Development Sdn Bhd, Maka Corporation Sdn Bhd, Pingat Harmoni Sdn Bhd, Wira Homeland Sdn Bhd, Haji Ali Group, Atira Development (M) Sdn Bhd and Sentosa Jaya Development Sdn Bhd. For example, SBJ Binajaya Development Sdn Bhd is the leading housing developer in Kelantan and has commenced business since 1976. SBJ has been involved in over 150 projects in different locations in Kelantan and has not stopped at building houses, but has set about transforming Kelantan into the premier financial and commercial state. Its housing projects are mainly concentrated in Kota Bharu and its vicinity. Besides that, SBJ also starts to build as many apartments as it can to meet the customer expectation. Its previous apartments are Condo Permai at Jalan Kuala Krai and Bayan Residency at Cherang. Incoming high-rise project is at Jalan Cemerlang, near Celcom Kota Bharu (SBJ Cemerlang Residences). More recently, SBJ also has jointly developed Taman Iman Jaya at Wakaf Che Yeh with Kelantan State Government.

Plenty micro homebuilders also provide contractual services to build a house on the private land such as Anjung Tulus Corporation Sdn Bhd, APPN Property and Development Sdn Bhd, Pelita Tujuh Construction, Assra Padu Sdn Bhd, DND Home & Construction, Dinda Resources, Rassco Builders Sdn Bhd, Quickstand Enterprise, DFA Home Builder Sdn Bhd and ZNH HomeLand Sdn Bhd. The approach is quite similar to "design and build" concept. Other small and medium housing developers in Kelantan include Almutawir Homes & Properties Sdn Bhd, Aman Realty Sdn Bhd, Asal Kurniaan Development (M) Sdn Bhd, AZ Inovatif Sdn Bhd, Berkat Makmur Jaya Development Sdn Bhd, Bridge Asia Development Sdn Bhd, DSI Consultant Sdn Bhd, Rushdan Yusoff Properties Sdn Bhd, Wadi El-Qura Development Sdn Bhd, Cleantech Development Sdn Bhd, KB 18 Development Sdn Bhd, Kayangan Property Sdn Bhd, Mentari Arus Sdn Bhd, SBS Indahjaya Development Sdn Bhd, SMS Costroltech Sdn Bhd, Reka Perdana Sdn Bhd, Rekajitu Development & Properties Sdn Bhd and AZ Refqee Home. For a record, Berkat Makmur Jaya Development Sdn Bhd, Rushdan Yusoff Properties Sdn Bhd and Wadi El-Qura Development Sdn Bhd also have residential projects in Terengganu even though in micro-scale. Essentially, their presence makes the competition for housing construction in Kelantan stiffer and more competitive.

In addition to local builders, a few non-local housing developers from Kuala Lumpur, especially, are in their way to develop some high-valued properties in Kota Bharu such as KEN Holdings Bhd and TH Properties Group. Plus, state-owned enterprises like Binaraya PKINK Sdn Bhd, SPP Development Sdn Bhd and Tunjong Development Sdn Bhd also vigorously undertake residential projects for Kelantanese. In general, housing entrepreneurs in Kelantan are considered creative in surviving their business as they will joint-venture their activities with other firms, create construction projects among peers, market their own products, become property agents and managers as well as offer real estate consultancy in order to earn considerable income for the sake of firm sustainability.

2.6.9 Terengganu

As a consequence of increasing population in Terengganu, many novel entrepreneurs have dabbled in housing industry. To name a few, Ting Mee Development (M) Sdn Bhd, Faktor Murni Development Sdn Bhd, Ikatan Unggul Bina Sdn Bhd, Ariz Construction Sdn Bhd, Warijitu Sdn Bhd are among the new players in this state. They are usually small in size with less than 10 permanent employees and build less than 10 units of houses annually. Similar to Kelantan, a number of micro homebuilders can also be traced in Terengganu. They include Aruni Bina Sdn Bhd, Anizam Unggul Sdn Bhd, MRA Properties & Development Sdn Bhd, Bumi Timur Properties Sdn Bhd, Success Global Development Sdn Bhd, WHA Properties Sdn Bhd and Zaraza Bina Sdn Bhd.

There are also some homebuilders that denote themselves as professional housing contractors. They are known as housing developers that involve in building contract for private homes such as Common Perspective Sdn Bhd, Air Mudik Sdn Bhd, Muhibbin Development Sdn Bhd, STMG Property Sdn Bhd, SuriaSabar Legacy Construction Sdn Bhd, Vista Infinity Development Sdn Bhd and Warisan Hartanah Sdn Bhd. Other notable residential housing developers in Terengganu include Konsortium Perumahan Rakyat Terengganu Sdn Bhd, Trigon Pawana Sdn Bhd (both are operated in Kuala Terengganu), Anjur Kekal Sdn Bhd (in Besut) and Jubimas Sdn Bhd (in Kemaman). At the same time, seasoned players who have successfully sustained their business through ups and downs are Pakadiri Developer Sdn Bhd, Syarikat Mohd Noor Yusof Sdn Bhd (SMNY), NPPR Resources Sdn Bhd, Syarikat Hassan Hussain Sdn Bhd, Tanjong Developer Sdn Bhd, DM Holding Sdn Bhd and Wan Z Bina Sdn Bhd.

As a leading developer in the East Coast region, pioneering the very first guarded community development in Kemaman, Terengganu, Prestimega Sdn Bhd truly stands out among its peers in the region. Its signature projects include East Coast Park – Bandar Putra and Taman Banggol Permai. The existence of non-local housing developers can be discerned lately. Operating from Klang Valley, residential developers such as Mewapuri Sdn Bhd (with project in Kijal) and Merchant Street Sdn Bhd (with project in Dungun) prefer to conduct their business through site offices in Terengganu.

Besides private housing developers who are controlling the development business in Terengganu, state agencies such as Perbadanan Memajukan Iktisad Negeri

Terengganu (PMINT) and PERKAYA Bina Sdn Bhd also get involved in the housing industry. They might engage external talents to inject more innovations in their projects, but the essence of their projects is still to cater the housing needs of local populace. Quite recently, the Terengganu Family Development Foundation (*Yayasan Pembangunan Keluarga Terengganu* [YPKT]) has collaborated with few private property developers to develop its lands. Smart partnerships with Total Merit Sdn Bhd (Johor-based developer), VST Group (Penang-based developer) and General Height Sdn Bhd (Pahang-based developer) in three different residential projects have lifted up the investment margin of YPKT. As the landowner, YPKT would receive certain amount of percentage from the value of the project for land provision.

2.6.10 Pahang

The following list is the set of active housing developers in Kuantan, Pahang: Greenlast Housing Development Sdn Bhd, Nilaitek (M) Sdn Bhd, Quek & Co Sdn Bhd, Chee Lean Construction Sdn Bhd, Wonderful Sense Sdn Bhd, Franky Land Sdn Bhd, Alam Tenggara Sdn Bhd, Tunas Manja Development & Construction Sdn Bhd, Macinda Sdn Bhd, Liew Sing Development Sdn Bhd, Jaya Bina Properties Sdn Bhd, Syarikat Perumahan Pacific Sdn Bhd and Ming Hooi Development Sdn Bhd. Established in 1966, Quek & Co Sdn Bhd has been a reputable property player and pioneer across Pahang. From the completion of its first project, Starlight Park in 1968, Quek & Co Sdn Bhd is poised to see Taman Kempadang Jaya, be handed over in 2011. Meanwhile, despite quite novice in real estate, Eastern City Development Sdn Bhd has housing projects in three states, i.e. Pahang (in Kuantan), Terengganu (in Kemaman) and Johor (in Pontian).

The second city after Kuantan is Temerloh. Dynaton Holdings Sdn Bhd, Twinview Construction Sdn Bhd, Panantara Sdn Bhd, Jia Eu Development Sdn Bhd, LCS Development Sdn Bhd, Mega Village Development Sdn Bhd and Quick Speed Developer Sdn Bhd are among the leading residential developers in Temerloh. Other emerging towns also have their own housing entrepreneurs. For instances, S.A.C. Properties Sdn Bhd (projects in Jerantut and Triang), Consolidated Construction Sdn Bhd (projects in Bera and Triang), Propel Land Sdn Bhd (projects in Triang), K.C.Goh Development Sdn Bhd (projects in Bera), Perumahan Khasiat Cergas Sdn Bhd (projects in Lipis), Jedex Realty Sdn Bhd, One Impian Group (both have projects in Jerantut), Teratai Kilauan Sdn Bhd and Amalina Holdings Sdn Bhd (both have projects in Raub).

In Cameron Highlands, a famous destination for highland tourism, non-Pahang homebuilders have monopolised the housing industry for quite some time. Among them are LBS Bina Group Berhad (Selangor-based), Fern Valley Development Sdn Bhd (Perak-based), Ringlet Development Sdn Bhd (KL-based), Fabulous Vantage Sdn Bhd (Perak-based) and Omniplus Development Sdn Bhd (Perak-based). A property developer from Klang Valley which is known as Tapak Group also owns several projects in Lipis such as Taman Telemin Indah, Taman Cheneras Indah and Taman Pertama. Although Tapak Group is a non-local homebuilder in Pahang, it has successfully set its foot in Pekan, Kuantan and Triang as well. Another firm from Klang Valley, PDMC Holdings Sdn Bhd, also owns projects in Temerloh and Bentong, while a niche developer from Selangor, GCS Infiniti Sdn Bhd, has few projects in Temerloh.

The prime partnership between government and private sector can be seen from the steps taken by the federal authority to hire the expertise of local homebuilders. In particular, Dinamik Budimas Sdn Bhd, Nilaitek (M) Sdn Bhd and Binaan Desjaya Sdn Bhd are among the housing contractors that were selected to build affordable homes for Pahang under the project of "Perumahan Rakyat 1 Malaysia" programme, initiated by the Malaysian government. Also known as PR1MA, this scheme was established to provide affordable homes in urban areas to middle-income households. On top of that, parastatals such as Perbadanan Kemajuan Ekonomi Pahang, Pasdec Holdings Berhad, RT Development & Engineering Sdn Bhd and LKPP Property Sdn Bhd also involve in residential development in Pahang. They are responsible for promoting and coordinating the economic development activities of the state.

2.6.11 Sarawak

Sarawak is experiencing an accelerated growth in key areas including Kuching, Miri and Bintulu, aided by the state's development plan, Sarawak Corridor of Renewable Energy (SCORE) to emerge as the next frontier in the local economy. The city of Kuching, being the capital and administrative centre, will remain the star performer in the Sarawak property market. Many more strata-titled properties have been launched in the Kuching built-up area and other subcentres in Kuching-Serian, Batu Kawa, Muara Tuang and Matang, where prices are more affordable. There are over 200 housing developers in Sarawak with most of them are concentrated in four big cities like Kuching, Sibu, Miri and Bintulu. Housing developers that can be considered as old-timers in Sarawak include Central Land Development Sdn Bhd, Chieng Chuang Construction Sdn Bhd, Chong Kia Hoi Realty Sdn Bhd, Hock Kui Development Sdn Bhd, Soon Hing Group, Jak Kwang Builders & Developers Sdn Bhd, Y.H.Wong Enterprise Sdn Bhd, Ibraco Berhad and Oricon Sdn Bhd.

For example, Ibraco Berhad preserves a long-standing reputation dated back to 1974 when Ibraco Group has been a leading pioneer property developer in Kuching, Sarawak. Now, Ibraco is known as an award-winning township developer with numerous accolades from local and global recognitions. Another instance is Soon Hing Group. Established in 1976, it is also associated with Soon Hing Hong Development Sdn Bhd, Sing Song Hing Construction Sdn Bhd, Janting Enterprise Sdn Bhd and Shibumi Development Sdn Bhd. Among their flagship projects include Blessed Garden, Taman Janting, Taman In Ling, Taman Hua Ling, Taman Tian Ling, Greenville and Harvard Jaya.

Undoubtedly, Kuching is the homes for many property developers. Petra Jaya Properties Sdn Bhd, Top Green Construction Sdn Bhd, Howasar Contracts Sdn Bhd, Advance Synergy Realty Sdn Bhd, Timber Land Group, CMS Property Development Sdn Bhd, Naim Land Sdn Bhd, Sin Hai Min Development Sdn Bhd, Spago Property Sdn Bhd, Custodev Sdn Bhd, Grahome Development Sdn Bhd, Hong Seng Construction (E.M.) Sdn Bhd, Imperco Sdn Bhd, M&D Enterprise Sdn Bhd, Nursinar Development Sdn Bhd, OBYU Realty Sdn Bhd, Nationlink Group, Regal Group, San Chin Realty Sdn Bhd, Sendayan Group, Coramax Sdn Bhd, Tedfar Development Sdn Bhd, Usaha Cendera Sdn Bhd, Wunsan Sdn Bhd, Eight Properties Development Sdn Bhd, Double Dynasty Holdings Sdn Bhd and Kalien Enterprise Sdn Bhd are among the real estate developers that are originally from the capital city of Sarawak.

In Kuching, vast growth in Batu Kawah, Matang, BDC and Tabuan Jaya is expected as population grows significantly since the last two decades. Serious urbanisation problems and critical traffic congestion are among the challenges faced by the populace of Kuching and its vicinity. To pave away for well-planned growth in the state capital, shifting development away from the city centre was imperative through policy of decentralising development. This approach will inadvertently allow well-planned development to spread out in Kuching. As an innovative home-builder, Spago Property Sdn Bhd has taken its wise step by focusing on high-end residences. It is indeed a prudent decision to build high-rise apartments or condominiums given severe land scarcity.

In other parts of Sarawak, the survival of housing developers is still in a good shape especially in big cities like Sibu, Miri and Bintulu. In Sibu, for instance, Upcojaya Construction & Engineering Sdn Bhd, Nice Realty Sdn Bhd, WTK Realty Sdn Bhd and Yemas Development Sdn Bhd are among the few names that are usually associated with property development. In addition, Borwak Sdn Bhd, Rich Venture Construction Sdn Bhd, Fabulous Enterprise Sdn Bhd, Green Summit Development Sdn Bhd, Min Sing Company Sdn Bhd, Y.H.Wong Enterprise Sdn Bhd, Shin Yang Group, Yong Lung Group, Oxrich Garden Sdn Bhd, Homelite Development Sdn Bhd and Homemas Development Sdn Bhd are among the other names of homebuilders that based in Miri, while in Bintulu, there are Sarawak Land (Kemena Park) Sdn Bhd, BBC Construction Sdn Bhd, Chieng Chuang Construction Sdn Bhd, Double Dynasty Holdings Sdn Bhd, Hong-Yet Development Sdn Bhd and Paling Group that are well-known to provide homes for people.

Moreover, very small numbers of housing developers can be elicited in other divisions of Sarawak such as C.K.K. Enterprise Sdn Bhd (in Serian), NPK Development Construction Sdn Bhd and Wynyard Construction Sdn Bhd (both are in Sarikei). Nonetheless, there are quite a lot of large developers that have branches in few places. Hock Peng Realty Sdn Bhd, for example, has projects in all over Sarawak such as in Kuching, Bintulu, Sibu, Serian, Samarahan and Mukah. Consequently, this company is able to joint venture with the state government's statutory bodies and local authorities to develop and successfully completed numerous development and commercial projects. Talking about government, Permodalan

ASSAR Sdn Bhd is one of the myriad parastatals of Sarawak that aggressively involves in property development.

High returns that may be achieved once the project is completed have attracted many Sarawakians to enter this challenging business. A new boutique property developer in Kuching, Instarmac Development Sdn Bhd, along with other young developers such as Elica Sdn Bhd, Green Summit Development Sdn Bhd, Kwang Tai Development Sdn Bhd, SOP Properties Sdn Bhd, Benta Mewah Sdn Bhd, 7th Avenue Development Sdn Bhd and Wynyard Construction Sdn Bhd are among the active and fast-rising developers in Sarawak despite their lack of experience. Elica Sdn Bhd, in particular, is a new development group that is currently making headlines for its debut project, The Republic in Kuching. This maiden property development project has proudly clinched the prestigious Best Residential Property Development award at the 2014 South East Asian Property Awards Malaysia.

2.6.12 Sabah

The property development industry in Sabah is just starting to become more competitive and exciting. The surge in migration to Sabah's capital, Kota Kinabalu, has brought an economic growth spurt to its satellite towns, attracting savvy businessmen seeking to cash in on the trend. While the buying interests still remain strong, buyers now are equipped with information, and they are more selective with products that appeal to them. This situation is similar to other states in Malaysia due to tremendous growth of ICT. Nonetheless, the pace of property development in Kota Kinabalu continues to be on the uptrend owing to its strategic location and status as the capital city of Sabah. Kota Kinabalu's hot spots such as Foh Sang, Damai, Luyang, Lido and Kobusak have become the primary locations for housing developers. In spite of that, the rising price of homes in city centre have prompted developers to launch more affordable houses in order to fulfil high demands of lower-income groups. Some of them have opted for areas such as Penampang and Menggatal, about 30 to 40 min away from the city.

In Kota Kinabalu, there were not many residential launches in 2014 but there are a few that are worth eyeing. Among them are the Harrington Suites by Interland Properties Sdn Bhd, Jesselton View by Bina Puri Holdings Bhd and Sutera Avenue Residences by Mah Sing Group Berhad. A leading independent real estate consultancy in Malaysia, Knight Frank, also notes two medium-cost property developments, namely, Greenfield Residences and Ashton Tower, by Greenfield City Sdn Bhd and SCP Inanam 2 Sdn Bhd, respectively. Other recent projects in 2016 include SkyVue Residence by North Borneo Land Avenue Sdn Bhd, E-Residence @ Telipok by Cahaya Hartamas Sdn Bhd and D'Ambience by Wah Mie Group. On a different note, high-rise developments could also be noted in Kota Kinabalu especially since 5 years ago. Two condominium projects that are set to change the landscape of Jalan Lintas are Lido Avenue by Mega City Development Sdn Bhd and Lido Four Seasons Residence by Kinsabina Group. Also, RA Concept Development Sdn Bhd is launching a condominium project, known as The Light Residences, in the Jalan Penampang bypass. However, the rise in supply and prices especially for apartments and condominiums over the past few quarters is a cause for concern despite the supply of high-rises in Sabah's capital has grown over 50% in the past 8 years.

Many experienced property developers exist in Sabah such as Delta Group, Borneo Pacific (Holdings) Sdn Bhd, Wisma Merdeka Group, K.T.I. Group, SinaLand (Sinar Pembangunan Sdn Bhd), JP Group and Kinsabina Group. They are imbued with solid foundation and are now diversified into several other business activities. Kinsabina Group, for instance, is a leading residential developer. Its most recent project is Casablanca Residence @ Off Jalan Kolombong through its subsidiary company, Yat Chung Development (M) Sdn Bhd. Another example of veteran players in Sabah is Borneo Development Corporation (Sabah) Sdn Bhd, also known by the appellation, BDC. Since 1971, BDC has completed numerous development projects in Kota Belud, Kota Kinabalu, Tawau, Sandakan and Lahad Datu. The projects comprised of residential, commercial, industrial and high-rise buildings.

Another big property developer in Sabah is The W Group. Apart from real estate, this large conglomerate has also ventured into construction, manufacturing, trading and services industry. They are the leading homebuilders especially for high-rise buildings. Among the housing subsidiaries are Eurotra Sdn Bhd and Herowang Sdn Bhd. One more renowned developer is Hap Seng Land Sdn Bhd who owns projects in big cities of Sabah, i.e. Kota Kinabalu, Sandakan, Tawau and Lahad Datu. Similar to Sarawak, a lot of housing developers prefer to have their head office in the capital city even though their projects are located in different districts. Take a look on Boomarange Sdn Bhd and Blossomaju Sdn Bhd (both firms have projects in Keningau), B-Paramount Development Sdn Bhd (project in Tuaran) and Excel Pave Sdn Bhd (project in Kudat) for a clear model. In short, there are umpteen residential developers in Kota Kinabalu such as Prima Land & Development Sdn Bhd, Remajaya Sdn Bhd, Central Million Properties Sdn Bhd, Docomo Development Sdn Bhd, Wong Chik Lim Holdings Sdn Bhd, Menggatal North Development Sdn Bhd, Nadi Properties Development Sdn Bhd, Poly Eastern Sdn Bhd, S3 M Development (Sabah) Sdn Bhd, Spurwin Development Sdn Bhd, Chang Cheng Group, WK Majujaya Sdn Bhd and DF Development Sdn Bhd.

Originally from Sandakan, Wah Mie Group owns a number of associated real estate firms such as Hajat Melangit Sdn Bhd, Ramindah Sdn Bhd, Wilakaya Sdn Bhd and Asterasia Sdn Bhd. Among their notable housing projects include Taman Rimba, Taman Bukit Sepangar and Taman Indah Permai in Kota Kinabalu, while Taman Indah Jaya, Taman Sejati, Taman Airport and Taman Sri Rimba are their remarkable residential projects in Sandakan. Other distinguished housing developers in Sandakan include Bumient Development Sdn Bhd, Liberty Chain Sdn Bhd, Sunrise Device Sdn Bhd and Malbumi Holdings Sdn Bhd. Meanwhile, in Tawau, among the distinguished residential developers are JP Group, Supremin Sdn Bhd, Masama-Winquest JV Sdn Bhd, Rasa Sierra Development Sdn Bhd and Unika Housing Sdn Bhd. For Labuan, another federal territory of Malaysia, situated off the coast of Sabah, a Sarawak-based contractor-cum-developer called Eden Empire Development Sdn Bhd seems to be quite active in that island.

Not many new housing developers emerge in Sabah these days. Perhaps, the inner situation of economy in Sabah is uncertain with vague prospects of future development despite high demands for houses. Among the eminent fledglings are VC Bumijaya Sdn Bhd, Legacy Group Development Sdn Bhd, MH Land Development Sdn Bhd and WK Majujaya Sdn Bhd. VC Bumijaya, for instance, is aggressively developing its modern township called Bandar Mingo in Beaufort. Another property developer, Sino Land Group, also has projects in Beaufort such as Taman Eramas, Taman Selagon and Taman Lumat. State agencies also involve in Sabah's development. In particular, Sabah Urban Development Corporation Sdn Bhd (SUDC) which is a full-fledged property development company, wholly owned by Sabah Economic Development Corporation (SEDCO), a government agency of the state of Sabah. Since its inception in 1972, SUDC has been involved in developing various residential projects ranging from medium- to high-end housing. Among its flagship projects include Kingfisher Park and Kingfisher Cove in Kota Kinabalu as well as Seri Kayangan and Seri Maya Apartments in Lok Kawi. Besides that, another statutory body named Sabah Housing and Town Development Authority also plays its role in providing low-cost houses and affordable homes to Sabahans. Among its significant projects include D'Sri Gayang (JV with Golden River Development Sdn Bhd) and Kallang Villa (in Tenom).

Further fascinating fact is public-listed companies from West Malaysia such as S P Setia Bhd, Mah Sing Group Bhd, IJM Land Bhd, SBC Corporation Berhad and Titijaya Land Berhad also have ventured into property development in Sabah. Most of their projects are located in the capital city of Sabah, i.e. Kota Kinabalu and its conurbations. Noticeably, quite a number of residential developers in Sabah are also involved in the development of commercial and industrial properties. This scenario is rarely happened in other states. Their involvement in erecting shop houses, shopping complexes and light industrial factories are merely to fulfil the demand of population as well as to diversify their business strategically.

Overall, only a small number of property developers have successfully ventured in both East Malaysia (Sabah and Sarawak) and West Malaysia (Peninsular Malaysia). They are, among others, Mah Sing Group Berhad, IJM Land Berhad, Hap Seng Land Sdn Bhd (the property division of Hap Seng Consolidated Berhad), Nusmetro Group, SCland Group, SCP Group, TPPT Sdn Bhd, Tropicana Corporation Berhad, Eng Han Property Sdn Bhd, Pembinaan Purcon Sdn Bhd, Samling Group, Amcorp Properties Berhad and Pristana Property Group. For illustration, Mah Sing's project in Sabah is located in Kota Kinabalu, IJM's project is in Sandakan, while Tawau is the preferred location of Hap Seng. Another examples are Samling Group, a boutique township developer with flagship projects, among others, Desa ParkCity (in Kuala Lumpur) and Marina ParkCity (in Miri, Sarawak) as well as Pristana Property Group, which have commenced their property business since 1980, with projects in Peninsular Malaysia, Sabah dan Sarawak, i.e. Pristana Garden in Kuching, Sarawak; Puncak Luyang Condominium in Kota Kinabalu, Sabah; and Aristana & Castana in Sungai Buloh, Selangor. Other excellent models of diversification are portrayed by Eng Han Property Sdn Bhd (with projects in Terengganu, Penang, Negeri Sembilan, Selangor, Pahang and Sabah) and Amcorp Properties

Berhad (with projects in Sibu, Sarawak and Shah Alam, Selangor, respectively). On a related note, despite its existing projects in Klang Valley, Negeri Sembilan and Johor, Pembinaan Purcon Sdn Bhd has also ventured in Lahad Datu, Sabah with its maiden project called i-Peak Developments that encompasses business centre and residences.

Nationally, another player in Malaysian housing industry is cooperative enterprises. The participation of housing cooperatives in developing housing estate for Malaysians is very limited and confined to few locations only. Nevertheless, there are a small amount of housing cooperatives that can be regarded as the pioneers and old-timers of the industry. They are Koperasi Belia Nasional Berhad, Koperasi Tunas Muda Sungai Ara Berhad, Koperasi Perumahan Melayu Perak Berhad and Koperasi Perumahan Malaysia Berhad. Again, with handsome returns expected from real estate development and investment, few other cooperative societies have decided to delve into property development. They are Koperasi Pasar Melayu Baling Berhad, Koperasi Kemajuan Tanah Negeri Johor Berhad, Koperasi Peserta-Peserta Rancangan FELCRA Seberang Perak Berhad, Koperasi Perumahan Angkatan Tentera Berhad, Koperasi Pembangunan Daerah Kemaman Berhad, Koperasi Koppes Sarawak Berhad and Koperasi Wawasan Malaysia Berhad. Albeit new in the arena, they have shown great potential to progress further. Take the example of Koperasi Wawasan Malaysia Berhad that has established Kowamas Development Sdn Bhd to be fervently involved in housing development industry in Kelantan.

2.7 Foreign Developers in Malaysia

As mentioned above, the bulk of the Malaysian real estate sector is overwhelmed with local property players. Over the last few years, however, given the strong growth potential of the real estate sector, the presence of foreign developers has been discernible. Traditionally, Malaysia has been attractive to the large property players from Singapore and large conglomerates like CapitaLand Limited has had inroads into the Malaysian property market as long ago as the 1990s, especially across the causeway in Johor Bahru. CapitaLand Malaysia, for example, is a developer, investor and manager of residential, commercial and integrated developments and is currently focused on the developments of high-end residential complexes in Kuala Lumpurs's prime residential areas like KL Sentral, Mont' Kiara, Bangsar, Kuchai Lama and Embassy Row. Developers of this ilk cater to the rarified and exclusive residential portfolios and development units are almost always small in numbers.

Foreign developers also tend to focus on high-growth areas like Penang and Johor where the market potential is stronger because of the proliferation of high-income individuals and families. These are considered mature markets, and the ever-presence of a large expatriate community provides better sales potential. Singapore's sovereign fund, Temasek Holdings, has gone into a joint venture with Penang Development Corporation to develop the Penang International Technology

Park in Batu Kawan, Seberang Perai, with a sizeable residential content. However, in January 2017, the said project has been deferred due to current property market conditions.

The ascension of China as a regional and global player over the last few years has also seen its effects in the Malaysian real estate. In the last 3 years, major Chinese developers have entered into joint ventures with private and state-owned firms in a prolific manner. Penang has seen the involvement of Consortium Zenith-BUCG for the Gurney reclamation, while Bandar Malaysia in Sungai Besi, Kuala Lumpur, sees the involvement of China-owned company China Railway Engineering Corporation (CREC) together with Iskandar Waterfront Holdings of Johor to develop what is touted as the largest real estate development in the country, a purpose-built integrated township with a gross development value of RM160 billion. Although property experts lament the possibility of supply overhang due to the current glut in commercial properties in and around Kuala Lumpur, the Chinese developers are undeterred as they see freehold land as a valuable commodity and are generally unfazed by economic conditions that they consider irrelevant with the belief that such conditions will eventually change for the better.

The involvement of Chinese developers is even more profound in the case of Forest City, located near Tanjung Kupang, close to the southern tip of Johor that sees the state government in joint venture with Country Garden Holdings, reputed to be China's third largest developer. This 1386 hectare development on reclaimed land will see the construction of 70,000 residential units with a gross development value of RM470 billion when completed in 2045. The unique marketing feature of the project, which appears to cater exclusively for Chinese nationals, has resulted in some controversial decisions especially with regard to the freehold status of the land and subsequently, the strata titles to be issued. As recently as March 2017, concerns were raised when it was learnt that the Chinese government has imposed capital controls on outflow of funds from China, thereby jeopardising the investments already made in good faith by Chinese nationals attracted to the Forest City allure.

2.8 Conclusion

Due to the economic uncertainty, Malaysia is going through a tough patch currently, but as far as the housing industry is concerned, the grass is still considered reasonably green here and will only draw more people to its pastures. Its generally stable economy has attracted many foreign investors to pursue land development projects in Malaysia such as Marimo Land Sdn Bhd (from Japan) for a project named O'hako@Puchong Jaya and Alfranko Development Sdn Bhd (from Russia) for a project called NOVO Ampang. More recently, two Singaporean-based companies named Macly Group and Roxy-Pacific Holdings Ltd have established a joint venture to initiate Macly Equity Sdn Bhd in order to build The Colony @ Infinitum that comprises 723 residential units as well as 31 retail units. These are quite apart from those already mentioned above and constitute a welcome relief to see excellent development projects emerged in the industry. Undoubtedly, developers whether local or foreign are becoming more thoughtful to homebuyers, as they take into consideration the demand for quality living and the impetus for greener environment. With the rise in demand, local residential developers need to capture the interest of new and seasoned property buyers with ever evolving design concepts, a variety of built-in amenities and very competitive prices. They have to heed the success factors of excellent housing peers for their business survivability and further growth. In the upcoming chapters, success stories of selected Malaysian housing developers will be further narrated.

References

Jaafar, M., & Ali, R. (2011). A study on indigenous housing developers in Malaysia. *African Journal of Business Management, 5*(16), 6891–6900.

Jaafar, M., & Sahari, M. H. (2013). The development of network linkages and housing developers. *International Journal of Sustainable Society, 5*(1), 78–96.

Shuid, S. (2015). The housing provision system in Malaysia. *Habitat International, 54*, 1–14.

Sirat, M., Che-Hamat, A. F., Abdul-Aziz, A. R., Rahim, A., Salleh, H., & Yaakob, U. (1999). *Low-cost housing in urban centres of Malaysia: Issues and challenges*. Penang, Malaysia: Penerbit USM.

Part II
Small- and Medium-Scale Private Housing Developers

Chapter 3
Maka Corporation Sdn Bhd

Keywords Maka Corporation Sdn Bhd • Wan Mohd Dzulkifli Wan Daud • REHDA Kelantan • Phasing strategy • Cash flow and Internet marketing

Maka Corporation Sdn Bhd is a Kelantan-based housing development company, not known as a big player in this industry, but rather a company that is sustainable and self-reliant as well as excels in financial management. From a micro firm that began with the construction of 104 units of low-cost houses in 1989, their group of companies have now completed some 26 housing projects, mainly in Kelantan and Terengganu. Due to the ability to well-manage their cash flow, they have successfully implemented all the projects from the beginning without resorting to any financial assistance from the bank.

The company is headed by Wan Mohd Dzulkifli Wan Daud who had served as Kelantan Branch Chairman of the Real Estate and Housing Developers' Association of Malaysia (REHDA) from 2000 to 2008. This former student of the Malay College Kuala Kangsar and the alumni of the University of Aston in Birmingham, United Kingdom, is hardly a recognisable figure in the public, but he is quite a well-known personality among industry players especially in Kelantan. For him, the success of the development company is not measured by how much profit the company makes or how large the size of project being implemented but on the capability to tactfully manage the business cash flows and to complete the projects successfully and on time. To sustain the business beyond his generation, all his children were educated in the disciplines that are required for the running of the business in property development, such as civil engineering, e-business and IT study. They were also given a 'hands-on' practical training to take over the management of the business through learning and experiencing the know-how of real estate intricacy.

Among the flagship projects of this company are Perumahan Maka Utama, Taman Asyraf, Taman Aqilah and Taman Putra Jaya. Interestingly, the Founder had also immortalised all his three children's names in his housing projects like Taman Desa Arif in Machang, Taman Desa Johan in Pasir Mas and Taman Desa Azam in Machang. Through a successful branding exercise, together with a sound track record of quality development and customer satisfaction, this company is poised to remain a key player in property development in Kelantan for many years to come, especially in providing much needed affordable houses to the citizenry.

M. Jaafar et al., *Business Sustainability Model for Malaysian Housing Developers*, DOI 10.1007/978-981-10-5266-8_3

3.1 How Did You Actually First Expose Yourself to the Business in the Housing Industry?

Prior to delving into this industry, I knew nothing about the profession of a housing developer. My interest since school days was actually to become a civil contractor, getting involved in the construction of buildings, road and drainage works as well as other related jobs. That was why I took a civil engineering course at the university. During my younger days, I saw a lot of people who were doing businesses in constructions were mostly wealthy even without higher education. Obviously, I was not really interested with a monthly-paid job especially working in the government which I felt at that time, during 1980s specifically, was quite slow going to my liking and unchallenging. I have always considered myself as an active person, creative and have my own thinking. Having such characters would not be very suitable to be working under someone else.

After returning from the United Kingdom in 1981, I started working with Sime Darby Plantation Bhd at Nova Scotia Palm Oil Mill in Teluk Intan, Perak, in the field of palm oil production. According to my Manager at that time, I was selected for the job for having high leadership qualities and very active in organising student activities during my university days. I used to serve as Secretary of the Malaysian and Singaporean Students' Association (MASSA) at the university. I was also the Organising Chairman of the All-England Malaysian and Singaporean Students Sports and Cultural Carnival that was held in Birmingham in 1979. After working for 5 years with Sime Darby Plantation, the last one being at Sungai Bogak in Bagan Serai, Perak, I felt rather like running on empty and unsecured because of not being able to do what I love, which is to run a business as a building contractor. During that time, many of my friends in Kelantan were doing well financially, working as a contractor even having studied lower than the SPM[1] level. In fact, many of my UK colleagues who worked with the government had their own side businesses as contractors, and they were doing well too! And from that moment on, I started looking for a new job in the field of construction.

In 1985, I had an offer to work as an Assistant Project Manager at a well-established housing development company in Kota Bharu, Kelantan. While working in that company, I found out that there is another more interesting career in construction than being a building contractor, which is housing developer. Although my salary was very much lower than my previous job at Sime Darby, I knew this appointment could provide me a great opportunity in my quest to pursue a career in entrepreneurship as a housing developer. I think, to become an entrepreneur, working in a small company is better than in a bigger conglomerate. We will have more opportunities to learn and experience all the required aspects of the business. With the big company, your exposure will be much limited to the position you hold, not unless if you are working at high level.

[1] SPM stands for 'Sijil Pelajaran Malaysia' (Malaysian Certificate of Education), a national examination conducted for all fifth-year secondary school students, which equivalent to eleventh grade in America's K-12 of education.

3.2 What Is So Interesting About the Career in Housing Development?

I think this career gives opportunity for ordinary people like me to become rich in just a short time. I came from an underprivileged family. My father at that time had just retired from working as a lorry driver with Public Works Department (JKR) in Tanah Merah and I have 13 siblings. I was the second son and the only one having a job since my eldest brother at that time was in the USA, completing his PhD degree. I needed enough money to support my younger brothers and sisters in their education. My parents' small house also needed extra rooms to be built too to accommodate the growing children.

Unlike contractors, housing developers do not need to depend on others such as government, politicians and etc to secure a job. A developer creates his own job by finding suitable lands to be developed, either through joint venture or by direct purchase. The revenue is earned from the sale of the houses they built. In general, developers do not need to be active in politics to succeed. Besides that, in the property industry, the risk of price depreciation does not happen in our country, especially in Kelantan. In fact, the price will always go up over time, and there is no expiry date for land, not like any other perishable goods. If one has sufficient knowledge and experience, doing business in this industry does not require a large capital to endeavour. In my opinion, an intensive and comprehensive working experience for 2 years is adequate to start a career as a small-scale housing developer. I actually was ready to start my own business after working for about 2 years, but due to economic recession facing Malaysia at that time, I had to postpone my intention to until 1989. But in the mean time, with consent from my employer, I started doing contracting jobs to build bungalows for government officers, as an effort to seek additional income.

3.3 Don't You Require a Large Capital to Become a Housing Developer?

When I first started in 1989, I did not have any money actually. I took RM5000 loan from my wife and together with one share partner, we ventured into development of our first housing project in Machang, Kelantan, which consists of 104 units low-cost houses, now known as Taman Desa Arif. The project was completed in 1992 without much problem. How did I do it? I can share the detail later, if you want.

If we do not have large capitals, we can start development through a joint venture agreement with the landowner, to avoid payment of land price. A 20:80 sharing of completed housing units is one of the systems that has normally been used by new developers in Kelantan. The construction part of the project can be subcontracted to building contractors with payment being made by instalment after receiving progress payments from the homebuyers. Professional fees to consultants such as architects, engineers and lawyers can be done the same way. Those days, even payments to the authorities can be made by instalment and with post-dated cheques.

All we developers have to do, apart from monitoring the quality of construction, is to ensure that the marketing is going smoothly as planned and the progress payments from the buyers are received on time. We do not want the construction to go too fast but the money to come in too slow. In other words, we must know and monitor the management of cash flow astutely.

To solve marketing problems, there are many ways of doing it. The developers can provide various facilities and incentives so as to attract buyers. Sales commission can be increased for the salesmen to put more effort. On top of that, our own sales promotion can be intensified.

When the houses are sold, make sure that the processes of the buyers' housing loans are managed efficiently with the lenders (the Malaysian Treasury or with local banks). We must have a practical and a working system to keep track of the movement of buyers' application files in order to make sure the progress payment can be released on time.

When the amount of cash-out can be offset with cash-in, the injection of development capital will be in the minimal level. In the event of deficit spending, part of the land from the project can be secured to obtain a financial loan from the bank.

3.4 Many People Said that This Joint Venture (JV) System Is Less Profitable, Compared to Direct Purchase of the Land from Landowners. What Are the Effects of These Transactions?

Let's say, a 1-acre piece of land is offered for cash sales with a price of RM150,000. But under a JV system, the landowner in Kelantan basically will ask for around 20–25% of the total units of houses built on that piece of land, which is approximately 12 units of terraced houses or 6 units of detached bungalow houses. So, from 12 units of single-storey terraced houses built by the developer, 3 units will be given to the landowner. If that terraced house is sold at a price of RM165,000 per unit, the total return the landowner will get will be RM495,000, that is RM345,000 or 230% higher than the cash price of the land. If single-storey three-bedroom bungalows are built, the landowner will normally get two completed units from the 1-acre land developed. If the house is sold at RM300,000 per unit, the landowner will get RM600,000, that is RM450,000 or 300% higher than the cash price of the land.

With such a high return paid to the landowner, the profit due to the developers will be severely affected. When the developers did not get a reasonable profit from the development, made worst with a poor and slow housing sale, the risk of uncontrolled cash flows will occur. This is actually the main cause of the abandonment of housing projects which will be further worsened if bank loans are involved.

Based on those factors, we normally are not too keen in developing project on a JV system. For us, it is better to acquire a land on a cash basis, even by using bank loans because the interest that will incur is only about 10% a year. Purchasing the already subdivided lands which are a little more expensive, but with instalment payments, is anytime better than acquiring land on JV basis. According to our experience, even if the land is purchased cash at RM150,000 per acre, with the present high cost of development and at the current sales price in Kelantan, the profit the developer can make out of that development is only minimal.

3.5 How Could Your Company Continue to Remain Active in This Business Despite All These Precarious Environments?

I think I am very fortunate to have made changes to my business strategy, specifically after the economic recession in 1998. Prior to that, our business approach and practices were usually done like any other local developers would do, especially the Bumiputera companies. We were so possessed to develop as many projects that we could, to ensure that our sign boards could be seen everywhere, although the acquired projects were not going to bring in much profit. If we had a piece of land, we would try to develop the land all in single phase, to make the project look huge, and we would be seen as big-time entrepreneurs. Although I managed to complete and sell all the houses that were built, I found that the profit earned from the project was not adequate enough to enable me to buy bigger land than the previously developed. The land prices began to rise and the ones who actually benefited most were the landowners surrounding the projects that I developed. Their land price rose up drastically without having to do anything. At that moment, I realised that my business strategy was all wrong, and we seriously needed some soul-searching to do.

One day, when REHDA Malaysia was having its National Council Meeting in Kota Bharu, our President at that time made me realised that even though I was professionally very capable, my business strategy and method were all very wrong. He advised me to buy a larger land (even through bank loan and any other methods), and if the land is smaller, do not develop it all at once but in phase basis with highest return as possible. And he also counselled me, if possible, all investments over the land should be recouped from the development of the first phase of the project, thereon the balance of the land will become free assets for the company.

Being aware of the fact, I consolidated all funds I had and tried to find a bigger land which is usually cheaper in terms of rate per acre. We also developed them wisely in phases and tried to make the most out of it. With ample stock of lands in hand, I would not be tempted to purchase any land offered not unless the price is really reasonable and will provide us significant benefits when it is developed later on.

3.6 Could You Please Share with Us Your Specific Managerial Practices?

To be honest, I actually never have taken any formal courses on management, be it financial, project or business management. My management knowledge was gained from my working experience with previous employers and from my long involvement with REHDA. Whenever I was requested to give executive talks or lectures on effective company management, I tried to share all the good management practices that I had gathered from those institutions. Those practices include proper control and monitoring systems, decision-making methods, operation strategies and management practices that are efficient, accountable and so on.

As the owner of the company, I need to have a system to enable me to keep track of things that I consider important to ensure that my business runs smoothly. I have created a simple system that allow me to monitor the performance of the housing sale, progress of each unit of the house being built, the status of buyer's loan application, the company's financial status and performance and overall project progress report. With the system, any potential problems could be identified easily, and measures to rectify them will be taken immediately.

I do not practise meeting system in running my business. If there is a decision to be made or any problems to be solved, I will only discuss with the persons concerned; face-to-face or by telephone. To improve management efficiency, all employees are assigned to multi-tasks jobs and all of them are involved in marketing which is directly under my supervision. For every successful sale, a handsome commission will be paid.

Although the implementation of the project is managed by a project manager, I am still very much involved in the quality control of the project, just to ensure they are built to the standard that I have set. For me, the main factor that will sustain marketing success and maintain good reputation of the company is the enduring quality of the house produced that met the customer's every expectation.

3.7 Do You Set any Particular Production Target to Be Achieved?

Absolutely. Any business that has no production target will record no growth in performance. I consider my company as a factory. Every year I will set the amount of products in terms of housing units that have to be produced. At the end of each year, although the year's production are still incomplete, when the new year starts, I will make sure the construction of the new target will commence, even during the economic downturn or even if we need to engage new sub-contractors to do the job. New target will usually exceed the previous one in terms of income. The target we set must be reasonable and financially manageable within our means, without having to seek financial assistance from the bank. We do not want to be too greedy, do we?

Lately, due to the difficulties in obtaining new land for development, I have tried to reduce the usage of my land reserve but still maintaining the target in terms of revenues and profit. It is made through upgrading the quality and the status of our houses such as from single-storey to double-storey houses. The housing design could also be enhanced, so that it can be sold at a higher price.

3.8 How Do You Deal with Public Relation and Networking?

A good public relation is based on the concept of respect towards each other. In dealing with the government and local authorities, I will try to handle them myself whenever I can, except for routine affairs, as a show of respect. I will also try to establish close relationship with them, possibly with officers at all levels, not just with heads of the department. I also try to distance myself from developing arrogance and 'showing off' as these kinds of attitude will drive people away. I found a lot of people love to help humble persons especially those they considered as successful. Based on my observation, people with knowledge are more respected than those who pretend to be rich and act smart. Do not be surprised; humility can get you a long way.

3.9 Can You Share Some of the Marketing Methods Practised by Your Firm?

At the early stage before the company was publicly known, marketing was done by participating in housing campaigns organised by REHDA Kelantan to target local buyers. To target those staying outside Kelantan, sale promotions were conducted in the locations where many Kelantanese would be easily found, such as at military camps and police quarters, all over the country. They were our main target because for them the opportunity to shop around was limited. With our attractive incentives offered such as no down payment imposed on the signing of the S&P agreement, their housing loan being processed and borne by the developer, sales would be usually successful if they were eligible for loan, even though the houses at that time had not been built yet. Sales to the local buyers were a bit more difficult because they preferred to wait until the house or the project was almost complete, then only they were ready to commit to the purchase.

We also practised MLM (multi-level marketing) or a pyramid style concept in our marketing. When any of our buyers came to my office, I would entertain them personally, to the extent that I was willing to send them to the bus station if necessary. When the relationship was established, I took the opportunity to encourage them to recommend our houses to their acquaintances. For every sale achieved, they would be rewarded with a handsome commission. Our target set for each of these buyers was to bring in at least two new prospective purchasers. I found this method of housing sale as the most effective of all at that time. Perhaps we need to use that method again for our future sales.

As time passes by, our marketing approach has changed, mostly due to myriad problems of abandoned projects after the era of economic recession in 1997. Homebuyers became more cautious and would only purchase house that is almost completed and from reputable developers only. As such, we had to act in accordance to this new development to ensure the successful marketing of our housing project. Having said that, we began to introduce a new sale strategy and approach, thus establishing a new concept of housing sale through our marketing tagline 'Perumahan Maka: Duduk Dulu, Baru Bayar' (directly translated as *Maka's Housing: Live First, Then Pay*). By this concept, we gave assurance to our purchasers that they would not be incurred any payment including loan repayment to the bank, not until the house is fully completed and ready for occupation, except if they are not eligible for 100% housing loan. To enhance confidence to the purchasers, they were encouraged to visit the project before making the purchase. More often than not, we would launch the sales only when the houses were nearly completed to ensure that the buyers are utterly satisfied when making their purchase.

To achieve maximum exposure for our products, we have embarked aggressive promotions through newspaper advertisement, promotional banners and Internet channels like Facebook and website. Internet marketing is the most effective and the cheapest way for spreading information to our potential buyers, especially to targets that are far beyond Kelantan. To create better confidence, the company's website was created and designed to look more professional and trustworthy. Any valuable information that can help enhancing company's image and branding is fully exposed through our official blog, http://makacorp.blogspot.com. I am happy that we have successfully achieved our marketing target all these years. Housing sale is no longer a challenge to us even if the country's economy is in a difficult situation.

3.10 What Are the Main Challenges Currently Facing the Industry?

The main problem is the difficulty in securing suitable land for development, at an affordable price. Of late, the land prices are getting too high in Kelantan. In Kota Bharu town area, the price of Malay Reserve Land has already reached as high as RM1 million per acre. A 5000 square feet vacant land for bungalows in the rural area has already reached RM50,000 to RM65,000 per lot. The price of building materials and labour charges are on the sharp rise as well. The housing price in Kelantan could not be increased to the level as in other advanced states such as Pahang or in the West Coast because the market in Kelantan is limited due to restrictions on land transfer and lower demand.

I would expect in the near future, a single storey house will no longer be built in Kelantan due to the factor of high land prices. Now, we do not see many housing projects being built, except for private houses built by the landowners themselves. The phenomenon of high-rise development has already begun in Kota Bharu and is expected to spread to other smaller towns very soon. I think the government need to address this issue seriously. The government should review their policy on land use in housing areas such as the size of road reserves, green space and etc. The rate of development density must also be re-examined.

Apart from that, they can also reduce the cost of development by loosening up conditions that requires sewage treatment plants (STP) to be built for projects consisting more than 30 units of houses. This policy is not feasible to be applied to the states like Kelantan because the price of houses here is too cheap to cover the cost of STP construction. They should also consider reducing the development contributions and charges for housing projects since all the infrastructure facilities are being provided by the developer. Because of these policies, my housing is getting smaller and smaller by the day. And licensed housing developers are getting lesser too. The government can also ease the burden of the project cash flows by allowing a few percent of housing units be sold to non-locals.

3.11 In Relation to That, What Are Regulations and Laws that Hinder the Growth of Housing Developers?

Frankly speaking, the most absurd regulation that was ever created by the government is the requirement for developers to deposit RM200,000 for issuance of each project's housing development licence, under the guise of preventing developers from abandoning the projects. Although this amount may be small to big-time developers, it is causing huge cash flow problem to small developers because it has to be paid upfront. And this ruling does not even help preventing a project from being abandoned. Besides that, much more regrettable is that this deposit will be increased again soon to an amount equivalent to 3% of the project development costs (excluding land price) and this money will be withheld for 24 months till after the date of completion of the project.

Due to this strange policy, many developers in Kelantan are beginning to build houses less than four units to prevent from having to apply for development's licence. As a result, the production of houses is now tremendously reduced and thus getting more expensive. These types of projects also do not require basic amenities (tar roads, drains and fire hydrants) to be provided. At the end, the main losers will be the homebuyers.

As far as I am concerned, the government should find another measure to solve the problems created by these unscrupulous developers, without causing problems to genuine developers and that will not affect the production of the houses. For one, the government could consider giving exemption to genuine developers with good track records from having to comply to these rulings. I think it is timely for the government to create a category of 'Master Developers' or '*Pemaju Berwibawa*' for this industry as similar to that created in the construction sector.

3.12 The Government Intends to Introduce 'Build-Then-Sell' System Very Soon. How Will that Impact You as a Developer?

For us in particular, that will not be much of a problem because our project implementation method now is almost similar to this approach. With the introduction of this new system, our housing production may decrease a bit but our competition would also be much reduced. But the most affected by this ruling will be the housing industry in Kelantan.

The lands in Kelantan are mostly Malay Reserve Land and small in size, and such a situation will not attract large developers to do their projects here. Being as such, the housing supply will be greatly reduced and the price will then increase. The government will not able to cover the shortage thus depriving the housing needs of the citizenry. A lot of jobs will be lost and local economy will be adversely affected. Honestly, I hope the government will look seriously at this scenario before implementing the policy. I cannot help thinking those people advocating this policy must be profoundly very stupid.

3.13 Have You Ever Had Problems in Getting Loans?

Our company has never applied any loan from the bank or any financial institution, for the purpose of operating our business or development, be it 'bridging loan' or 'business loan'. We only helped to arrange 'end-financing' loan for our homebuyers. When the Central Bank of Malaysia (BNM) tightened the condition for 'end-financing' approval in January 2012 by reducing the margin of loan repayment amount to 60% of the monthly disposable income of the purchaser, many of our potential buyers from the private sectors found it hard to comply, thus affecting our sales quite a bit.

For developers outside Kelantan, the impact on the housing sales may not be very obvious as the income of the population is high and the market is vast. In Kelantan, the population general income is lower and the sales are only restricted to Kelantan Malays. This BNM regulation really gives a severe impact to Kelantan developers, who are mostly Bumiputera. We really hope that BNM could be more considerate in drawing policies that accommodate local conditions especially to those houses built on Malay Reserve Land that have limited market.

3.14 Business-Wise, What Was the Toughest Challenge Ever Faced by You?

Probably the toughest challenge was during the economic crisis of 1998, my first recession experience after 9 years in business. There was an interesting incident occurred during that time. I was developing one housing project in Rantau Panjang through one company I set up for the purpose of helping a brother by taking him as a share partner. Due to restricted fund available, I invited other family members to invest in the project with a guaranteed repayment profit of 30% per year. Many had agreed to join and invested.

Even though the project went rather smoothly, we were having sales problem. When the first phase was completed, we managed to sell only 10% of the houses. I felt very depressed. I could not understand why such a beautiful project, located in a very strategic location and sold at an attractive price, still found it hard to get buyers. Investors were starting to press for repayment of their investments plus their dividends.

After a long and deep thought, I believed that such predicament could have occurred due to the element of 'riba' (usury), which is forbidden in Islam, being practised in dealing with that project. I straight away went to see a friend at Bank Islam in Kota Bharu and asked about the Islamic method of implementing the profit sharing concept of doing business (the Mudharabah). After I fully understood about the concept, I called the investors and seek their agreement to convert the original agreement to Mudharabah, with a profit sharing of 40:60. All of them agreed. One week later, by coincidence or not, all the houses were fully sold and the investors got their money back plus profits as promised.

3.15 Many Fellow Players Opine that You Do Not Fully Utilise Your Potential to Further Develop Your Business. How Do You Respond to This?

This happened because I am not the type of person who likes to take risk. All strategic actions I took were based on the level of knowledge and experience that I have and on my financial capabilities. Basically, I do not enjoy having debts and if possible I will try not to take bank loans. In this kind of situation, it is very difficult for a company to grow faster, I presume. Yes, my company did grow, but at a very manageable phase.

I was also offered by several big companies to do joint ventures with them, but I refused nicely due to our different business philosophy. I also tried to diversify into the business of exporting frozen sea products to South Korea and in the stock market. But both of them failed miserably because I did not have enough experience in the field and just relied on others to manage them for me. The lesson learnt from these episodes is that we should not dabble in businesses that we are not good at and do not rely on others to manage our money.

3.16 If You Dare Not to Take Risks, How Could You Manage to Earn More Profits?

The business in the housing development is very risky and strictly controlled by rules and regulations. It is also an industry that involves a large amount of finance. It does not allow us space to make mistakes. If we fail in one project, it would probably pull other projects all together. Once fallen, it is extremely hard to recover. Therefore, to really succeed in this industry, we have to be smart. We must know the exact time when to go fast and when to go slow. We must understand about economic cycles that usually occurred in our country because when recession comes, the chance to fail is high if we are not prepared.

I cannot afford to fail because many of my family members depended and worked for me, either as contractors or at the office. Also, I need to take care of my children and wife. If I fail, all of them will lose their sources of income. On top of that, I do not have connections at higher levels that can assist me when I am in trouble. I just cannot fail come what may. That is why I always try to play by the book and complete my project properly. Well, I love to make more money like other people do but at what price?

3.17 What Else Do You Practise to Sustain Well in Business?

Doing business in housing development is just one of the many ways to make an honest living. Being a Muslim, my belief is that all the bounties come from God. We can only make an effort, work hard and strive to the best we possibly can, but eventually it all depends on His provision and blessing. Being convinced by that fact, I tried to live and do business by following His guidelines; do what Allah enjoins and leave what are forbidden, always be good to parents, stay away from 'riba' (usury), be generous in charity and perform the zakat (tithe). Last but not least, we should be grateful for everything that was given to us.

We were also been taught to recite certain prayers when we are in need of His help to ease our hardships. We were told to distance ourselves from developing arrogance and from people with characters and attitudes that were so despised by God, such as showing off, untrustworthy, sinners and those who like to beg. After all the efforts have been poured to the best that I possibly could, I will leave to God to do the rest.

3.18 As a Conclusion, What Is Your Most Commendable or Proudest Achievement?

I do not think I can be considered successful just yet. But I am grateful with what God has given me and that He did not test me hard enough that will make me fails half way. My biggest satisfaction or the proudest achievement, I think, was to be able to delight and make my parents happy.

My mother must be pleased to see a beautiful *madrasah* (religious school) named after her, 'Madrasah Hajjah Che Esah', built at Taman Maka Utama, Pasir Mas. I believe my late father too would be proud to see that a housing estate in his hometown, Tanah Merah, was named after him (Perumahan Haji Wan Daud) if he were still alive. That 5-acre land of his, abandoned and idled, was developed into 13 units of bungalows and 37 units of terraced houses. As a result of the development, that retired JKR lorry driver managed to feel the taste of money and was able to spend on anything he wanted, without having to ask from his children.

Those were the most satisfying moments achieved during my long years in property development, the pleasure of making my old parents happy and successfully fulfilling a father's ambitions before he died. And as for me, there is nothing else more that I want: *No desire for more and no more to desire.*

Chapter 4
Wira Homeland Sdn Bhd

Keywords Wira Homeland Sdn Bhd • Hanafi Bin Mamat • Land surveyor • Current economic situation • Demographic study and staff motivation

Wira Homeland Sdn Bhd was founded in 2000 as a class A civil contractor. Since its establishment, Wira Homeland has successfully developed few housing projects in Kelantan. Some of the completed projects in Kelantan are Taman Alia Putri, Taman Nurul Hana, Taman Bayu Wira, Taman Apam Permai and Taman Tengku Aminah. All the said projects are located at Pasir Mas, Kelantan.

Datuk Hanafi Bin Mamat is the man behind all those achievements. As the Founder of the company, he always spends his entire life making sure that Wira Homeland one day will become a key player in housing development business. He started his career as a surveyor at the Department of Survey. Then, he tried to contest in the general election in 1990 as requested specially from the then UMNO President, Tun Dr. Mahathir Mohamad. But his destiny said politics is not his future; then with the technical knowledge that he had, he tried to start this business. From an ordinary surveyor at the Department of Survey until he becomes one of the successful Bumiputera status developers, there are a lot of things that we can learn from his successful story.

4.1 Datuk Hanafi, Could You Please Share Your Background in Property Development and What Motivates You to Delve into This Field?

I finished my study, Diploma in Surveying, at UTM (Universiti Teknologi Malaysia) and then continued my degree at North East London Polytechnic, United Kingdom. I started my career at the Department of Survey of Pahang and then moved to Terengganu before I moved to Kelantan. At Kelantan, I was given the responsibility to manage and supervise four districts, namely, Pasir Mas, Tumpat, Tanah Merah and Jeli (used to be the small province). Suddenly, around 1990, I received a letter

M. Jaafar et al., *Business Sustainability Model for Malaysian Housing Developers*, DOI 10.1007/978-981-10-5266-8_4

from our former prime minister, Tun Dr. Mahathir Mohamad, who requested me to contest in general election as Barisan Nasional candidate. So, I resigned with a 24-h notice and gave my full attention and focus to win the election. Luck was not with me and I lost the battle. But that is not the ending of my life. With my experience in surveying, I started my surveying firm named 'Wira Ukur Consultant' on October 1990 and then dabble in land development on 1999 until what you can see now. What motivates me to jump into land development is due to my vast experience in subdivision, partition and engineering survey. So what I plan is to synergise all of them into one business called land development from buying land bank, subdividing it and then developing it.

4.2 May You Please Share with Us How Do You Develop and Manage Your Company and the Challenges that You Have Faced?

When I was in the government, I knew so many information on land and the market price. I usually bought lands from the persons who applied for government land and wanted to sell it. The price was reasonable at that time, around RM30,000 to RM40,000 per acre. That was how I initially started this business. I started by employing one manager to manage my company. His job scope was more to operation of the company and also marketing. My company advantage was we already got approval for site division and subdivision earlier. I still remember we started our project with the land size of 5 acres (equivalent to 14 lots per acre). Then, gradually we expand our business by rolling our profit. I presume our land bank can last for 5 years from now. My focus for development is at areas covering Pasir Mas, Kota Bharu, Tumpat and Tanah Merah. The strategy is based on our study on demographic factor. Most of the young Kelantanese professionals live at Klang Valley, so our target customers are more to government servant and private sector.

To do business in Kelantan is a challenge. When I first started this business, land price was not that high and we still have many Malay Reserve Lands. Unfortunately, no big developer comes from Bumiputera companies. The existing developers do not have capabilities to run big project. Politics for me is not a problem. During my starting period (in 1999), PAS[1] was the ruling party in this state, but I got no problem working with the government. The officers are all professional. Furthermore, I only develop private land with no government land involved.

One more thing is the culture in Kelantan as we apply sub-labour system whereas we buy all the equipments and subcontract the works to others. But we must be particular in supervising, so that we can make sure they deliver the job on time. As

[1] PAS stands for 'Parti Islam Se-Malaysia' (Pan-Malaysian Islamic Party). As one of the most influential movements in Malaysia, this party often attempts to place Islam more consciously at the forefront of its campaign policy. PAS is part of the opposition alliance known as 'Pakatan Rakyat' (People's Pact).

for now, we can see from the statistics issued by the National Property Information Centre (NAPIC) that Kelantan is the second last or the last in take-up rates. So, it is my hope to see that one day, many job opportunities will transpire and the purchasing power of Kelantanese will become higher.

4.3 How Do You Manage Your Life as a Land Developer and How Do You Manage Your Time? Also, What Is Your Future Planning for Your Company?

Honestly, I would prefer if people recognise me as a licenced land surveyor. I sent my two sons to study on surveying, respectively, at London and Newcastle. One of them is pursuing his study in Master's degree and his brother had just finished his professional examination. That has proved how I love surveying; I persuade both of my sons to study in this field. Hopefully, once I leave the politics arena, I will focus more on land development. I have to balance my life since I have two companies, surveying and development firms. Even if the challenges are really tough, I am blessed to have a very cooperative team working with me. I employ one engineer, one manager and other supporting staff. I also hire one general manager to oversee my land development company. Back to when I first started my company, I put in my mind to that one day my land companies could hire more people and own as many land banks as we can. I also hope that population could increase tremendously and their purchasing power is high. Indirectly, it will give benefits to land developers like us, as many of them will buy properties here. There is a trend that we can see in Kelantan; people here like to buy land and then build their home by themselves. That is why during the economic recession in 1998, they can survive by selling back their land. They came to developers like us and applied subdivision.

4.4 What Types of Innovation that You Have Made?

To be frank, nothing much I have done. This is due to so many rules and regulations for us to comply with. We deal with so many government departments and agencies. Maybe what I can say is that we make some innovations with regard to the design and layout, according to market needs. In terms of marketing and promotion, we go for several ways of 'offline and online' such as direct selling, mass media, flyers and private invitation to potential buyers whom we got their names from various associations and unions. I admit that private invitation is very tedious as many administrative stuff need to be done. We also have our own website that we always promote our projects there.

4.5 Can You Share the Current Scenario in Kelantan in Terms of Housing Industry? Please Give Advice to Newcomers

My advice is to be sensitive with the current economy situation. I ask you, who are going to buy houses? For sure, young professionals and those who just started their career right? So logically, if one location, for example, Kelantan, got many job opportunities, the demand should be high. If not, it will be stagnant. Based on the statistical data, purchasing power in Kelantan is not so good, and many of young generations prefer to work at Klang Valley, so it is a bit difficult to sell more houses here compared to other states, except for Kota Bharu.

In addition, banks nowadays tend to tighten their loan terms and regulations resulting in not so many people can make a loan. So, it will affect this industry. Banks require many documents such as salary slip and other supplementary documents. Most Kelantanese do not have their payslip, so it is difficult for them to proceed with the housing loan. With regard to Chinese community, they would prefer to buy from non-Bumiputera developers, unless we can offer strategic development location, and the projects have commercial value.

For Kelantanese, the acceptable housing price is between RM250,000 and RM300,000. Some areas in Kota Bharu can go up to RM400,000 for bungalow. Location is their main consideration, whereby design is not that important for them as long as there is a façade with 3–4 rooms per unit. I just want to stress that land price is increasing here; sometimes it can go up to RM70,000 per lot, but price per housing lot cannot go more than RM300,000. That is the challenge here.

4.6 Does Economic Situation Affect Your Firm Planning?

Yes, it does. Take the scenario in the year 2000–2005; during that time there were so many government projects at Kelantan such as schools' development, water supply to rural area and so on. Most of these projects were developed by big contractors from class A, B and C. While they were busy with these big projects, it gave chance to us to do land development without their interventions as they did not disturb our business. It was an advantage to land developers like us. But now, there is a tendency for small-scale developers to enter this market. They do not need licence to do the development as they only build houses less than four units per project. This is an advantage to them as they can offer price that we cannot afford. This one is the current challenge that we are confronting right now. Above all, as a small new company, we did well to weather the recession.

4.7 What Are the Personal Traits that You Deem Have Influenced Your Business Survivability in This Industry?

First is resilience and second is courage and not to forget hardworking. This field is challenging, so there are many stakeholders to take care of. You must be very particular with your target, and do accurate calculation for both money in and money out. For me, I always put minimum target as a benchmark with no maximum target. Higher achievement will be better and I assume that as a bonus.

My philosophy is we do business and give back to society. I always dream to have a business so that I can create more jobs for my people as well as uphold my people's dignity. It is parallel with the intention of UMNO[8], a political party that I join. That is why for me, profit comes number two. Big profit does not make me happy, but contribution, zakat (tithe payments) and taxes that I pay from my revenue do.

Furthermore, to be relevant in this field, you must have strong leadership character. So many politicians do this business because they are capable, but in the end, not many survive. They may be a good leader, but they do not have a strong team to support them. Symbiosis between leader and team is a must; leader should give accurate direction while team should be able to execute it.

Lastly, your very important asset is your staff. Usually, we motivate them by giving yearly bonus. Either we make profit or not, I try to give them bonus every year. There was one time during recession period that I got no job for 7 months. What I did was I asked my staff to sow chillies on our land bank. In the end, there were some reserves for us. Yes, not so much, but at least there is something than nothing. This scenario is just to assure them that I appreciate them in order to keep them motivated. So, our staff will be loyal to us.

4.8 Could You Please Share with Us Your Valuable Experience in This Industry?

The economic recession during 1997 really taught me a hard way. One vital thing I learnt is never ignore small business. During that time, we cannot rely much on big players such as MRCB (Malaysian Resources Corporation Berhad), SEDC (State Economic Development Corporation) and other big companies as they did not make payment as jobs delivered. So during that time, the only hope is small businesses such as individuals that come to us and sell their lot or sublot. That is how I survive during recession. I always remind my staff not to neglect these small people. Treat them as similar as you treat big players.

4.9 Any Last Words from You to Be Shared with the Public?

Even though I already resigned from politics, even before I went to London, I am still wishing for Malay survival. As a developer at Malay state, Kelantan, how I wish that one day, among all the Bumiputera developers in Kelantan, there must be few from them that can be outstanding and at least at par with non-Bumiputera developers such as SBJ Binajaya Development Sdn Bhd. I suggest that we do paradigm shift, to think that nothing comes easily. Working hard and determination are vital for Malay developers to survive. One more thing, at our best level, try to protect Malay rights especially on the issue of Malay Reserve Land. I can see nowadays that many of us tend to sell their land as the price is skyrocketing. Thirty percent quota for non-Bumiputeras as they required, for me is not applicable in Malay states like Kelantan and Terengganu as for now.

Chapter 5
Pakadiri Developer Sdn Bhd

Keywords Pakadiri Developer Sdn Bhd • Hussin Omar • Prior experience • Land bank • Joint venture • Staff benefit • Network and relationship

Pakadiri Developer Sdn Bhd is a low-profile housing development firm that perhaps only locals of Terengganu might know of. Starting out in 1988 as Pakadiri Sdn Bhd, this family-run business began as a company shared between the current Managing Director, Mr. Hussin Omar, and his friend. Combining a management degree with over 30 years of experience in the construction business, he brings astute direction and management to the table. Assisted by his loyal General Manager, Mr. Hasan Adam, since 1989, they have steered the business to build up an impressive record of successfully completing every housing development project. While not as glamorous as other fellow established developers, their road to success was paved with similar kinds of challenges.

Mr. Hussin Omar is a quite unassuming person at first sight, yet friendly once chatting commenced. Perseverance and swift adaptability brought the desired success 7 years after he started his first company when he registered Pakadiri Developer Sdn Bhd in 15 November 1995. He personally believes that no amount of success comes without a measure of adversity but admits that his journey was 'a steep hill to climb'. Company-wise, they own swathes of unused land that can be converted for the use of property development. Another interesting fact lies on their projects' names that usually start with 'kalongan', as a brand or trandmark of the company, to reminisce the root of their first project originally initiated at Teluk Kalong, Kemaman. Hitherto, they have completed almost 20 housing projects with the notable ones include Taman Kalongan Impian, Taman Kalongan Ilmu, Taman Kalongan Melor and Taman Kalongan Murni. Prior to that, among the projects under Pakadiri Sdn Bhd are Taman Seri Kalongan (Phases 1, 2 and 3) and Taman Kalongan Intan in Dungun which was the last project. Now, they are moving a step upward through a plan to build the Open University campus in Terengganu.

Housing is not just their business – it is in their blood. At least, that is the impression one gets when speaking with Mr. Hussin and his team. Their housing's strong success was certainly not achieved overnight. It took a lot of endurance, patience and thorough planning. Known to tread very carefully in their business despite changing times, the team is willing to share the key factors to their success. A valu-

M. Jaafar et al., *Business Sustainability Model for Malaysian Housing Developers*, DOI 10.1007/978-981-10-5266-8_5

able chance to speak with this affable yet tenacious property magnate of Terengganu has shed some light on this often misunderstood industry and shares how he recovered from a disaster that nearly wiped off everything he had put in the business – and thrived. For a record, he currently acts as the housing Chairman of *Dewan Perniagaan Melayu Malaysia* – Kuala Lumpur Malay Chamber of Commerce, Malaysia (Terengganu Branch) and due to his contributions to the state development, he was conferred the dignitary title of Darjah Setia Sultan Mizan Zainal Abidin Terengganu.

5.1 Please Tell Us About Your Family Background?

Back then, my father was a *batik*[1] maker while my mother was a *songket*[1] weaver. At first, my parents did not encourage me to be involved in land purchases, but I still bought lands using my savings. Those days, land was quite cheap. Now, after more than 30 years, I sometimes think that it was a waste to buy lands without a proper plan as I had missed huge opportunities to develop them since the 1980s.

After my graduation, I worked as a Marketing Manager in Sime Darby and was in charge for the eastern region as I had to monitor Pahang, Terengganu and Kelantan. Then, I started working with Pesaka Sdn Bhd. During my tenure there, I established Permint Plywood Sdn Bhd and a few other subsidiaries. If I am not mistaken, in the early 1980s, *Perbadanan Memajukan Iktisad Negeri Terengganu* (Terengganu State Economic Development Corporation) had granted forest reserves of about 250,000 acres under the management of *Kumpulan Pengurusan Kayu-Kayan Terengganu* (Terengganu Timber Management Group) to be coordinated with Permint Plywood Sdn Bhd, which was under my care.

5.2 Prior to Setting Up this Business, What Did You Do? What Are The Factors That Prompted You to Delve into This Industry?

I worked with Pesaka Terengganu for 10 years. In 1985, after I left that corporation, I secured some engraving works from the Central Bank of Malaysia, Employees' Provident Fund (EPF) and other government agencies. But before that, I worked under Lin Deco and Craft Sdn Bhd for a while, as they received a RM2 million contract from Istana Iskandariah (the official residence of the Perak Sultanate) in Kuala Kangsar. Thereafter, in order to diversify our services, I formed Pakadiri Sdn Bhd with my friend. Since then, we decided to join the housing industry. However, since my partner was competent in engraving skills, having worked with the Ministry of Tourism and Culture, we decided to break off the partnership. I took this

[1] 'Batik' and 'songket' are types of traditional fabric from Indonesia and Malaysia, recognised as handicraft products of both nations.

chance to take care the housing development part, Pakadiri Sdn Bhd, while he took over the wood carving. This restructuring process occurred around 1988. So, with a capital of RM250,000, we started our housing project. Since the market was uncertain, we built low-cost houses, admittedly with bad financial management. For that reason, I needed to sell some of my lands to pay off the project costs; during that time, I took a bus rather than used my car to visit the project site. This situation continued for about a year and, in 1989, Mr. Hasan Adam joined this firm. I had the opportunity to perform *hajj* (an annual Islamic pilgrimage) in Makkah that year, and I asked Mr. Hasan Adam to try selling all the houses, and fortunately, he managed to realise and fulfil my wish as the project was sold out.

We then proceeded with our next projects. When Pakadiri Sdn Bhd received a government project, I decided to open a new firm, Pakadiri Developer Sdn Bhd, to manage small-scale projects. Both firms are involved in housing development and they complement each other for the purposes of survival. Between 1987 and 1988, there were about 15 housing developers who were actively involved in the industry, but we are proud to say that we are the only one who survived till now. During the early stage, we focused on low-cost houses since high-end properties were so hard to sell. Also, the government had provided subsidies in terms of cement and other bureaucratic benefits for the developers to construct low-cost housing. They expedited approval for subdivision and housing plan, as well as offered rebates on stamp duties and legal fees. Those were the incentives for Malaysian housing developers to build low-cost houses in the late 1980s and early 1990s.

5.3 How Did You Manage to Acquire Quite Sizeable Land Banks in Terengganu?

In my case, I loved to buy lands since my younger days, officially beginning in 1974. I had finished my tertiary education in 1972, and in 1974, I bought a piece of land in Kampung Sungai Penchala with a price of RM28,000, which may now cost up to RM3 million to purchase. I then bought another land in Kerteh, to be exact in Kemasek, near the oil region of PETRONAS. I bought it at RM1,000 per acre, with a total of 6 1/2 acres. Now, the price of that land is rising up to RM100,000 per acre. I am still holding on to it and have no intention to sell. These activities happened mostly in the 1980s, during my service with Pesaka Sdn Bhd.

You may be surprised with my ability to accumulate huge land banks in this state, right? Actually, I got them from various channels. Some were through acquaintances that introduced potential sellers or landowners who wished to sell their land to me or through my own initiative to seek land with good potential. Unlike some people, I used to spend my leisure time riding a motorcycle around villages looking for land with potential for good development purposes. Now however, having made our name locally, several landowners and brokers will approach me to negotiate on potential land deals. With such abundance of land banks, I anticipate building hotels and bungalow projects in the near future.

5.4 Besides that, Are There Other Strategies Used to Accumulate New Land Banks?

It is our normal practice to buy new lands as replacement, once a certain project is completed. Consequently, the size of our land banks is increasing from time to time. At the moment, I have about RM60 million in terms of land values. We are also pleased to accept offers for JV development of housing projects. We have collaborated with Perbadanan Memajukan Iktisad Negeri Terengganu, in the sense of joint venturing strategy, for housing projects. With my previous record and credibility as one of their former management staff, there is no problem for them to approve the JV proposal as they trust in me.

Noticeably, the demand on housing is growing, while the supply is stagnating or even diminishing. Therefore, I anticipate that my children who will take over this business will be able to expand further and utilise the accumulated land banks to undertake sustainable development. This long-term approach is very beneficial to sustain the business in tandem with good managerial skills that I have nurtured on them through higher education and guidance.

5.5 What Were the Problems Faced by Your Company During the First Project?

During our first project, we were stucked at the first phase as the cash flow was very bad. We developed 50 houses for the first phase, 56 units for the second stage and 103 units for the third segment. On top of that, our project location was situated in a remote area, a little secluded from the nearest town, making the sales response to be relatively slow. Luckily, due to rigorous development of multinational factories in Teluk Kalong, Kemaman, such as Perwaja and Tioxide, we managed to improve the rate of sales. Once the problems of the first phase were settled, the flow of development improved and got progressively better and smoother. The best experience was when we built the second phase; we offered the buyers to add another RM4,500–5,000 for extra renovation, from the original price of RM25,000 per unit, and 50 buyers who were mostly Perwaja workers were able or qualified to get financial loan from one bank only, i.e. Maybank. That scenario had assisted in providing us with good cash flow.

5.6 In Order to Finance Your First Project, Did You Use Your Own Money or Bank Loan?

For the first phase that comprised 30 units, we secured RM150,000 from Bank Islam, while for the subsequent phases, we survived only with end financing with no bridging loans involved. However, for our project named Taman Kalongan Mas that consists of 113 units of houses, we applied for bank loan to assist us. For the time being, we also utilised a bank loan facility of RM2.5 million that acts as a 'standby fund', which we will use if the situation demands.

5.7 From Your Hindsight, What Are the Principles You Hold to Ensure that the Firm Keeps Moving Forward?

I always remind my subordinates to preserve quality materials, especially the wood. We also need to be tolerant with the buyers, perhaps in terms of price and services. If the buyers were late in payment, no additional charges were imposed. We adopt this approach to protect our good relationship with them. In fact, we do not practise the 10% deposit down payment and only require them to prepare RM5,000 for agreement fee.

I also stress on my staff to pay all our debts first. We need to prioritise our expenses by highlighting the most important one as we need to gain the trust from our suppliers and banks. All these things entail the element of integrity in our life so that everyone will be pleased. Apart from that, we need to be honest with our job, the product delivery must be on time, and it should be accompanied together with the Certificate of Completion and Compliance (CCC), in order to safeguard our reputation.

5.8 In Terms of Human Resource Management, Are There Any Specific Practices You Would Like to Highlight? Who Is in Charge of the Technical Part?

While Mr. Hasan Adam focuses on the administrative section, my sons, Farhan and Hazlami who are both university graduates, take care of the project at the construction site. Before that, we outsourced the services of draughtsman and architects, while for contractors, we usually engage similar people since they are familiar with our style of works. I would love to mention here that both employer and employee should conduct their tasks with high trust and accountability. Until now, monthly wages were always issued on time although the project is slow and sluggish during the start-up period, to uphold best practice of the working environment. We should also heed the importance of staff benefits and welfare by keeping our staff trained in order to work hard for us as they are our core asset. In the meantime, as an entrepreneur, we have to be hardworking as well, and try to maintain good relationships with our workers. When we visit the construction site, we do not need to act like a 'towkay' or become bossy in front of our subordinates. We need to interact with the construction workers in a pleasant manner and visit the site regularly to monitor the development progress.

5.9 What Are the Reasons that Made You Stay for More than 28 Years in This Industry? Is It Because of Your Interest or Other Factors?

Interest is one thing. The second one is because this job is based entirely on our plan, without the need to follow others' instructions. Let say, if we have a land to be developed, we only need to set the sales target and plan in accordance with the projected schedule. If we are a contractor, perhaps we need to lobby other people to get projects which leads to uncertainties, while this never happens in the case of housing developers. I acknowledge as well that I self-learn a lot from my experience working in Pesaka Sdn Bhd. Just to share with you, if we encounter any problems, we need to sit down and think wisely and try to figure out the solutions to the predicaments. If possible, try to prepare early and readily.

5.10 Talking About Targets, How Do You Implement Your Firm's Management in Order to Achieve the Preset Goals?

We inherently convene a weekly management meeting, or at least every fortnight, involving all senior management except my wife, although she is the main shareholder. During the meeting, if I were around, my children tended to be reluctant to give their opinions, so when I am outstation, I will inform Mr. Hasan Adam to let my children make the decisions after thorough and clear discussions. Hasan and I monitor every aspects of the firm's administration. My first son, Farhan, is entrusted to handle the construction part; Hazlami is on planning, while Nurul Amilin is in charge on marketing. I teach them business management with a gentle yet firm discipline. It is our practice to set the target a year ahead and try to accomplish it before the end of the year.

5.11 How Does This Firm Market Its Product?

Well, it comes from different channels and angles. We normally market our houses through local newspapers, brochures and billboards. Besides that, we employ Internet marketing and open booths at shopping malls that are able to attract crowds. For online marketing, we engage with *mudah.my*, and the responses are quite impressive. Out of ten buyers, four of them are from *mudah.my*. Having said that, I perceive that signboards at the project sites also play a vital role to impart information of our houses to the potential buyers. From the display, they will recommend our houses to their friends and relatives via word of mouth. Above all, I believe that buyers also look at our track records. For your information, in Terengganu, home-buyers will only give full commitment once the project has achieved 40% in terms of progress.

5.12 What Do You Consider the Most Unique Strategy Implemented by Your Firm that Distinguishes You from Other Fellow Developers?

We are proud to tell you that we are among the first housing developers to embrace the concept of 'build then sell' in this nation. This policy entails no claims via the progress of the works as we can only sell the houses once they are completed. To that extent, we think that we should build 20 houses first and then sell them, within a small-scale development concept through phase-by-phase strategy. After that, we will continue with another 20 houses and it applies to every project. This scheme is deemed pertinent to avoid issues related to liability period or project delays borne by the housing developers and project abandonment borne by house buyers. It means that once they purchase our houses and the loan is approved, we will hand over the house key. It is however true that the price will go up if we exercise the 'build-then-sell' model, but we will adjust the price in relation to the current market. On average, it entails a 10% increment of differences before and after the construction.

5.13 This Firm Is Considered a Veteran in the Industry; Have You Implemented Any Types of Innovation on Your Products?

Since we are focusing in the suburban areas, we are quite retrogressive in innovative applications. We try to adopt a little bit, but it is not compatible with our small-scale system, and our workers here are also not very familiar with current changes of technology. Nonetheless, the design still follows the current trend and transformation along with the materials used, since we are now utilising competitive and quality ingredients, commensurating with costs obviously. We are however still reviewing the novel idea of innovation to add value to our houses, mostly towards our bungalow projects.

5.14 Can We Say that You Are Focusing in Terengganu Because You Are Well Versed and Originally from Here? So, Do You Find It Easy to Be 'in Contact' with the Local Authority?

Absolutely right, because I know a lot of people in my state, and it is easier for me to deal as I have created a wide network here. For example, to get the approval for subdivision, I will get it done in a speedy manner. Again, since I have been working with the state government previously, I find the facilities to be very convenient as I have friends from the government servants that are eager to help me. Indeed, our dealings with the local authorities are positive, and I dare mention that the success of Pakadiri Developer Sdn Bhd is due to our amiable cooperation with various government agencies such as district council, town planning department, land offices and other related authorities. We also created good relationships with building professionals such as architects, suppliers, lawyers and bankers. In fact, we received a credit limit of 2 months from a number of material suppliers.

5.15 How About Your Managerial and Planning Skills? Are There Any Specific Systems Executed in Your Daily Business Affairs?

No explicit procedures to be shared. But we are presently practising the method of subcontracting regarding materials and labours where we feel comfortable with. One contractor, who has been with us for quite some time, performs substantially excellent work, as they are already aware about the quality that we want. The same goes for our suppliers; if we are confronted with some problems, we will be frank and they usually tolerate with us since they trust us based on our preceding records.

5.16 May You Please Share with Us the Housing Industry Scenario in Terengganu?

In this state, the scenario is quite different. Most of our buyers are government servants, so we do not see any effects of macroeconomics. Anyhow, we are still alert on current issues of price escalation and the downturn of the global economy. The competition here is quite stiff since small housing developers who build less than four houses are mushrooming all over Terengganu. For the record, if we build four units or less, we are not obliged to apply for the licence that will cost you up to RM250,000.

5.17 Last But Not Least, Do You Mind Sharing Any Words of Wisdom as a Golden Rule for Nascent Entrepreneurs Who Wish to Be Involved in this Industry?

Please bear in mind that once you enter this industry, you need to learn many things. For every effort that we pour in, we need to set certain targets to be attained and strive for them with hard work. Marketing should be strong, cash flows should be in good condition, and the planning must be proper and meticulous. On the other hand, I would personally call upon all the retirees to try this field; gaining RM5000 a month will be adequate for your life through housing development. Maybe they will just need to build and sell one house a year, and it really will suffice their elderly needs, of that I am sure.

Chapter 6
Macinda Sdn Bhd

Keywords Macinda Sdn Bhd • Jonathan Kok Keng Fai • Attention to details • Diversification • Project location • Product innovation

Macinda Sdn Bhd is a reputable property developer, based in Kuantan, Pahang, with more than 40 years of experience and was incorporated in 1976. Apart from being a developer and contractor that specialises in building, the group is also a renowned authorised Honda dealer in Kuantan, known as Macinda Auto Sdn Bhd. Its Founder and current Chairman, Mr. Lee Kee Hoo, believes on the principles of integrity and honesty by providing consistent quality and setting standards that are rare in the industry. A dialogue with its Executive Director, Mr. Jonathan Kok Keng Fai, who is the immediate past Chairman of REHDA (Pahang branch) and also the son-in-law of the Founder, has opened the essence of business success that is particularly unique to this company.

6.1 Can You Share a Bit About the History of Macinda and What It Stands for?

Let me give a bit history about this company. Macinda stands for Malay, Chinese and Indian. During its inception, we had the late Tan Sri Yahaya Ahmad, one Indian gentleman and Mr. Lee Kee Hoo as the shareholders. It is in line with the 1Malaysia concept, right? After that, the late Tan Sri withdrew due to his commitments with DRB-HICOM and other businesses. It was followed by our Indian partner a few years later. Now, this company is belonged totally to the Chinese partner after the Malay and Indian shareholders left the company.

M. Jaafar et al., *Business Sustainability Model for Malaysian Housing Developers*, DOI 10.1007/978-981-10-5266-8_6

6.2 Can You Share Your Company's Vision and Mission?

We do not have specific vision or mission statements. But we have targets that we always aim towards. We are committed to becoming a reputable developer with integrity. We are now on the right track and working very hard to achieve that. We want to develop properties that homebuyers are proud to buy or own. We promise to minimise defects in every project and ensure that rectifications will be made even after the liability period finishes. That is our commitment.

We also want to build something to keep for longer periods. So far, our projects are for sale, and none of them are kept for the future. We would love to have that kind of projects in the near future like hotels, condominiums and malls as they tend to be very sustainable. We consider this idea as diversification since we cannot depend on only one business nowadays. That is why we started the Honda business recently. Do not put your eggs in one basket only.

6.3 Who Actually Initiated This Firm?

My father-in-law, Mr. Lee Kee Hoo, started the company. During the economic downturn around 1997/1998, we decided that apart from being a contractor, we wanted to venture into the property development business. As a developer, you have more control of your business. You do not work for others but others work for you. By right, you get the money, but as a contractor, you are at the mercy of employers. Construction alone is thus very risky, and the returns are uncertain and tight. Due to that reason, we diversified our business into property development where we construct our own projects. In conclusion, instead of working for others, it is better to do it on your own.

6.4 So, the Development Part Is Headed by You?

My sister-in-law, Ms. Lee Hui Jing, is heading the firm now. She is the Managing Director while I am the Executive Director. I used to be the Managing Director before Macinda diversified the business as a Honda dealer.

6.5 How Does Macinda Start Property Development Projects and How Big Are They?

Our projects are not as big as you see in Kuala Lumpur, Johor Bahru or Penang. They develop townships; we have not ventured yet into projects of this type. You must have sufficient funds to do that kind of projects because development is a capital-intensive industry. Adequate funds are important especially when projects are faced with problems or the economy slows down; you still have to pay the contractors, or else your projects will be abandoned. Also, you cannot now simply claim your funds from buyers as the rules have become stricter. So, we build projects that we are confident of completing even if the sales are slow. To be able to do that, we must have proper planning, not just in design, quality and marketability but also financially.

6.6 Can You Share Your Experience in Handling the Development of Projects?

For us, whether you have two or three projects, or as many as ten projects, you still have to manage and handle those projects carefully. You still need site supervisors, sales team and so on. So, it is better to go for bigger projects as you still need to undergo the approval process from land acquisition and design drawings from the consultants until actual development. I started my development project in 2000 by totally subcontracting the works to contractors. I did that because of my virtually zero technical knowledge of this business. Everyday, I went to the site and supervised the construction and I slowly learnt from that. For my second project, I just outsourced the labour work, the rest I did on my own. After a while, we started to know which contractors are the best and we only worked with them. Now, my brother-in-law, Mr. Thomas Ting Wei Huat, who is a civil engineer, is the building contractor, and it is easy for us to ensure that the materials and workmanship are all up to standard.

6.7 Government Is Implementing BTS (Build-Then-Sell) Model Right Now. Will You Implement It on Your Future Projects?

Due to the current status, I would say 'not at the moment'. If you are talking about developers, you must have land banks first, either you buy it or do joint ventures. Land prices are hiking now and landowners will always ask for security deposits, about 30–50% of the land value. It depends on market forces and somehow on bargaining power as well. Let us say we have ten developers aiming at one particular piece of land; for sure the landowners have the power to mark up the price. That is why we would not prefer BTS, as the initial cost is already high. For us, we go for both, in that we have our own land banks and joint venture. Our projects cover areas of Kuantan, Pekan, Jengka, Sri Jaya and Maran.

6.8 Mr. Jonathan, What Are the Aspects or Features that Should Be Highlighted by Housing Developers?

To be a developer, you must have integrity and not to forget good track record. There are few developers that always set up a new company for every new project. I consider them with no track record. This is important, as customers always evaluate us from the track record. As homebuyers, I advise you to study the developer's track record first. It is also better to buy property during its launching time before the price goes up. I always remind my company to go into every detail for each project we develop, especially on the orientation of the houses. Houses facing west are not preferred as it will get hotter during afternoons. It is better to be facing north or south, or at least facing east, as morning sunrise is not that hot. These are the things that we need to take into account. At one time, we were among the pioneers who developed four-room terrace houses, and I can foresee with this additional room, homebuyers would certainly love it. We were also the earliest developers to build zero-lot bungalows in Kuantan. Now, people tend to buy a complete house, meaning that they do not have to perform any renovation after buying the house. To come up with all these innovations, we have to survey buyers' needs.

6.9 Any Successful Business Person You Admire Most?

There are a few persons, some of whom I do not know personally. I really admire one of Honda's sifu, no need for me to mention his name. I really respect his mentality, commitment and dedication as a business leader. The other one is a friend of mine, the visionary Founder and owner of Jaya 33 and Plaza 33. His very creative idea was to develop one building just for data storage. What I want to say was his thinking was well beyond his time, and he knows what is important for the future and so he decided to do that. These are the kind of successful persons that I admire. Last but not least is my father-in-law, Mr. Lee Kee Hoo, who founded Macinda without much experience and paper qualifications, yet he took risks and managed to build up our group of companies to where we are now with his "never say die" and "must always try your best" attitude. Even after retiring some years ago, he is always there for us to consult him and be our guide.

6.10 The Government's Rules Are Now Getting Stricter than Before. How Do You Adapt to This Circumstance?

We understand the government does that to protect buyers' rights. One of them is the low-cost property regulation. The Pahang state government instructs all developers whose projects are more than 5 acres to build at least 30% low- and middle-cost houses, and the prices must not be above RM80,000. We basically plan around with that; if say we have an 8 or 10 acres piece of land, we build or develop it in subdivided phases with each phase being less than 5 acres. We are then not obligated to build the 30% low-medium types of houses. The other one is regarding the Bumiputera quota. For your knowledge, there is no Bumiputera lots mentioned in the National Land Code, only Malay Reserve Land. The Bumiputera lot is something from the state governments. The problem is we are only allowed to sell those lots to Bumiputeras, and Bumiputera buyers also are not allowed to resell to non-Bumiputeras. That is why even Bumiputera buyers are not interested to buy Bumiputera lots. For us, this is unfair and not reflective of the open market. We thus come out with the solution; we have no choice but to increase the price of non-Bumiputera lots slightly to subsidise the Bumiputera lots. Whatever it is, we still follow the regulation as it is not our intention to disregard government requirements. We however need to ensure profitability so that we can roll our money for the next project. Recently, the government gave us a choice; they allowed us to release the Bumiputera lots to non-Bumiputeras, but we have to pay the 5% price difference to the government.

6.11 How Do You Market Your Products/Projects?

Do not compete with others in terms of price, but compete on other things such as location and innovation. Nowadays, customers prefer to have packages or gifts along with their purchase. Our products usually come with free add-ons like kitchen table tops, auto gate and air-conditioning points besides gifts like umbrellas, recycle bags and towels with our logo on it. Also, attention to details is something we need to focus on to be successful. For example, we make it easier for our customers by helping them to identify and fit in every room keys for them as well as a manual on the materials used like paint colour, roof and tile codes plus how to operate the fittings in the property. We consider this simple gesture as our practical innovation and our concern to the customers and hope buyers will appreciate the effort and benefit from this. If you can ensure that you are taking care of the small things, they will be confident that you will also take care of the big issues. In return, buyers get a good impression as we show our concern to translate small inconveniences into customer benefits.

I also want to add more on Japanese work ethics and culture. They are good in giving attention to details and that is why they can move forward. I try to implement this culture in all of our companies. In reality, we want to learn things that have values for us to further improve our lives.

6.12 Who Is Your Target Marketing Group and What Is Your Current Project?

It depends on location. Currently, we are developing houses near the airport. From our surveys, we found that people in this area can accommodate houses that cost around RM250,000. So, after considering the distance between their workplace and home, we decided to go for terrace houses with a price of RM230,000.

6.13 How About the Housing Scenario in Kuantan, and Are There Any Competition Around?

Yes, what is business without competition, right? The difference is competition here is not as severe as in Kuala Lumpur. Try to do something different, so that you can sell your products at a high price. As I mentioned before, we were the first developer to develop four-room terrace houses, enabling us to sell more, with more profits as well. The same goes with zero-lot housing; none of the developers here have started this type of project as yet. So, if we were to initiate this, we can sell better and at a higher price. This is the strategy to compete. People will go for better design, better features and better offers. We have to improve and innovate to add value. Auto gates and underground drainage systems are among some of the innovations that we have made. We did not launch our projects on paper; rather we launched them with show houses as I want to prove that what you see is what you get. Our practice is to build two show houses, one being fully furnished and the other is unfurnished. We need to keep on improving in line with the Japanese word 'kaizen', meaning continuous improvement. Technology is moving fast, and we must be not left behind in order to outperform others. During good times, try to build expensive and sophisticated houses, but during bad times, it is adequate to revert back to basics in order to survive.

6.14 Did You Ever Receive Any Awards from the Authority?

No. For me, I would prefer the humble and sincere recognition from customers as the best rewards. The best compliments are when the customers tell others about your good efforts and merits. The customers' satisfaction is the biggest award that you can get.

6.15 What Is Your Advice to New Developers?

Back to the initial topic, the developers have to be far sighted. They have to be decisive on what types of development they want to do according to the value of land. For example, we acquire one piece of land and it is quite valuable. So, we decide to develop a zero-lot project after doing some studies on the marketability. That is why planning at the beginning is very important. If you fail to plan, you are planning to fail actually.

Chapter 7
Dynaton Holdings Sdn Bhd

Keywords Dynaton Holdings Sdn Bhd • Alan Lim Kim Hin • Dare to fail • Quality product • Cost projection • Loyalty

Dynaton Holdings Sdn Bhd was incorporated in Malaysia on 21 May 1996. The company's core businesses are residential and commercial property development. This firm also undertakes all kinds of civil construction works, such as earthworks and sewerage piping and treatment plant, drainage and building constructions, and provides project consultation services. The Founder, Mr. Alan Lim Kim Hin, is currently leading the business together with his wife and partner to accomplish their dreams and mission simultaneously.

The company is formed in connection with a diversified group of associated companies, which has undertaken various developments and constructions managed by Dynaton Management Sdn Bhd and built by Temerloh Home Builder Sdn Bhd, with both companies certified for ISO 9001:2008 in March 2012. Dynaton Management is certified in managing development projects, while Temerloh Home Builder is certified for building and construction works.

With the experienced and dedicated workforce from the group of associated companies, Dynaton is confident with the total commitment, strength and capability to make all the present and future projects a great success. Its vision is to be the best and trusted mega township developer in Pahang. In the future, they are aspired to achieve the highest quality in all development and construction works by striving towards zero complaints on delivery and quality as well as maximising land value for landowner. Apart from developing and selling properties with high quality at competitive price, its Managing Director, Mr. Alan Lim, is envisaging to give property buyer peace of mind by providing designs that best meet their requirements, ensuring quality material and workmanship and completion on time.

M. Jaafar et al., *Business Sustainability Model for Malaysian Housing Developers*, DOI 10.1007/978-981-10-5266-8_7

7.1 Can You Tell Us About Your Educational Background?

My highest education was Form 6. At 32, I went to Sedaya College and took business administration, studying part time. Initially, I had preferred to take civil engineering, but due to the advice from Sedaya College, I eventually took business administration as I can learn how to manage my company later. It took me three and a half years to complete my study.

7.2 What Did You Do After Schooling?

I did everything under the sun. Most importantly, none of them relates to what I am doing right now. I used to work at a karaoke club, amusement centre as well as used car company. At that time, if I were to set up a business, I would rather open a workshop or sell used cars, not something big like my present company.

7.3 Who Motivated You to Start This Business?

This is my very own brainchild. I started in 1996 but was idle until 2004 due to the recession in 1997. My initial plan was to run this business by my own, but my family opposed. My brother-in-law suggested to me to take his brother-in-law as partner since he has experience in construction. I agreed, but I wanted to be the major shareholder because I firmly believed that I can eventually run this business. My share was 60:40 at that time. Now, I can say I am the only one running this business; my partner comes once a month to monitor the progress.

7.4 What Makes You to Get Involved in This Business?

I had intended to run a business, but before that, I was thinking what kind of business the richest man does? I categorised them by group. Printing business or selling fruits cannot make you the richest. I used to be involved in the timber industry but failed. Finally, I delved into development after taking few considerations. My father was a hawker before opening a grocery shop and finally involved in the rubber trade. Indeed, I have no background in the development business; I started from zero and learnt from experience.

7.5 What Is Your Business Philosophy?

Be honest and hardworking. Profit is not everything in business; good quality service is. With these two traits, I believe you will succeed since you will get trust and confidence from the public.

7.6 Do You Have Any Business Mentor?

God is my mentor; He always helps me.

7.7 Are You a Member of Any Developer-Related Association?

I am a member of REHDA and Chinese Chamber of Commerce and Industry, and I am also the President of *Persatuan Usahawan Muda Malaysia*–Young Entrepreneurs Association of Malaysia (PUMM), Temerloh Chapter. I was offered a Datukship, but I refused to accept.

7.8 What Was Your First Development Project?

My first project was a mixed project, which comprised of double-storey, single-storey, low-cost and semi-D types of houses. In addition, we had also four units of bungalow during that time. The land size was 10 acres for my first project. I started by developing phase by phase; it took a while so that people will slowly realise our development there.

7.9 What Is Your Company's Focus Right Now? Housing or Commercial Development?

Currently, there are some commercial projects which are under development. We will start housing developments again by next year. Pusat Perniagaan Chengal was our commercial project, and its development was commenced in June 2011. Next is Taman Saujana Jaya, a residential development project that was commenced 2 months later. Taman Cengal is now under construction, and we are doing initial work for phase 3 of Bandar Bukit Angin, while phase 1 is already completed and phase 2 is still under construction. The 100-acre township of Bandar Bukit Angin project is our largest project so far and will take 10 years to be completed.

7.10 How Does Dynaton Acquire Land for Development?

For my first project, I tried to approach my family, but they refused to join me as they did not believe I can do it. So I started by doing joint venture with outsiders. I did my work by giving my fullest effort and interest towards my job, and then my family started to put their belief in me. Finally, I acquired all the land pieces of my nine siblings either by joint venture or by buying from them. At the end, I managed to fulfil my promise to them, to successfully develop a housing project. Now, we acquire lands from various channels, for example, broker, joint ventures and individuals that want to sell their land to us. From our record, we have never joint ventured with the Pahang state government, as things stand at this moment.

7.11 What Is Your Target Group?

Due to rapid development at Temerloh, people from Kuala Lumpur and Kuantan are attracted to buy properties here. It causes the rental price to escalate in comparison with Mentakab and Kuantan. The rental price is about RM450 to RM550 for a house priced within the range of RM180,000 to RM220,000. For your information, Temerloh is the second town after Kuantan in terms of development. You will be able to see a lot of travellers from Kuala Lumpur to Kuantan dropping by here.

7.12 Who Is In Charge of Project Department and Who Is Responsible for Managing Sites?

We employ three project managers to look over site activities. I will organise weekly meetings on Mondays to gain updates from them and discuss the weekly planning. As the main contractor, we ourselves have our own labours, about 50 of them that we employ from Construction Labour Exchange Centre Berhad (CLAB), and they get paid biweekly. There is also another project wherein we outsource the labours from others, but we make sure that they deliver the best as ours.

7.13 Have You Applied Any Financial Assistance from Financial Institutions?

I did apply but the rule at that time was they will only pay you after you reach the breakeven point. So, I had to forgo the intention of borrowing money, except for the one time when we did apply for bridging loan for our first project. I then increased my marketing efforts by doing door-to-door sales. I asked my friends to support me and recommend my products to others. There was one time I chased whoever wanted to buy a house and recommended mine. Thinking back about all those moments, it seems impossible for me to be where I am now. Nowadays, we are in line with LHDN (Inland Revenue Board of Malaysia) requirements that need all developers to prepare annual financial forecasts, so from that we can carefully plan and spend our cash flow. It is important for us to prepare project accounts for every company that we are involved in. We revise it twice a year and I personally monitor all those accounts.

7.14 How Does Dynaton Handle Financial Matters and Quality Inspection?

As normally practised by a development firm that has many subsidiaries, we employ one purchasing officer to handle all purchasing-related matters. We regularly do quality checks on materials purchased by other officers, at the office and also at the site.

7.15 Temerloh Is a Small Town with Less Population Compared to Other Districts in Pahang, So How Confident Are You that You Can Excel Doing Development Business There?

As I mentioned before, Temerloh's location is very strategic, and even the population is more compared to Jengka, Chenor and Lanchang. It is located in the main road that connects Kuala Lumpur and Kuantan. Furthermore, most of the government's branch offices are still here even though some have moved to Mentakab. Mentakab is known as a Chinese settlement, while the Malays are predominantly in Temerloh. If you take Temerloh and Mentakab as your targeted business location, then it would be balanced. In fact, the distance between both districts is just 12 km. Also, we are talking about developing 100 units project only, not 1000 units like in Kuala Lumpur.

7.16 How Many Developers Are There in Temerloh?

When I started this business, there were about 30 developers in Temerloh, and now the number has increased to 40. The most active is Panantara, while Quick Speed and Twinview at Mentakab are catching up right now. I like this competitive scenario. We are the first to sell single-storey houses with the price tag of more than

RM100,000, double-storey houses for more than RM150,000 and semi-D houses for more than RM200,000. The reason behind this circumstance is that we cannot make profit if we sell with prices lower than that. But to attract buyers to buy with those prices, we have upgraded our houses and would not compromise on the quality. That is why I believe people will eventually buy our houses. Another thing I would like to add is we provide two parking lots per house. Hence, people considering buying properties with me realise it is worth the value.

7.17 How Do You Market Your Products?

For housing development, our target is government servants, and we do promotion at supermarkets, carnivals and public areas. For commercial development, we use a different approach since our targets are investors and businessmen, and 90% of them are Chinese buyers. Most of them come from Kuantan and Kuala Lumpur, and they are repeat customers. Some of them have relationships with previous buyers like relatives or friends. We used to ask iProperty for our Internet marketing, but there has been no response until now. Perhaps, they are very busy or our location does not interest them. We also have a Facebook account, but this channel is not contributing much to our sales. Furthermore, we did receive offers from some agents to do marketing and sales on our behalf, but no fix agreement has been made. However, I think their outcome is not so encouraging as well.

7.18 Why Do You Think Customers Prefer Your Products Compared to Others?

They are amazed by our quality, design and unique concept that suits their taste. Location also plays an important role to attract their interest. Plus, we offer our loyal customers extra discounts if they want to buy other properties from another project.

7.19 How Do You Come Out with Innovative Ideas?

I have my own habit. I would not buy medium-quality shoes, either I buy the cheapest one or go for the most expensive. The cheapest is temporary while waiting to buy the expensive one. The same goes for other things; always go for the best. Maybe because of that I always get new ideas on how to improvise and be innovative on new things. Additionally, I used to travel to Australia and get some ideas from there.

7.20 How Did This Company Create a Good Reputation?

Always be attentive to complaints from buyers. Once complaints are received, we promptly respond to it and rectify the problem soonest. We try to maintain a high quality as we want this principle to become our trademark from now on.

7.21 What Aspects of Quality Are You Really Concerned About?

As a main contractor, our main concerns are on quality. I am very particular, especially on materials used for development. We also utilise things above the specification levels required. I also go to the site frequently to supervise the progress of my projects.

7.22 How Frequent Do You Hold Your Management Meetings?

Once a month or on-demand basis since all the staff know their job scope already.

7.23 Can You Briefly Explain Your Company's Relationship with Stakeholders and How You Manage Them?

In terms of sub-contractors, we only deal with one contractor. This is important since we can develop loyalty and trust between them and us. We apply similar approaches when we engage with legal firms. They are willing to give their full commitment towards our work because we treat them amenably until they feel that their role is special and pertinent.

Local authorities or any government agencies are the most important entities as our stakeholders. We never had any problem with them; some of our proposals were rejected due to clashes with planning requirements. So, we amend accordingly and approval will be finally achieved. On a separate note, we have mutual understanding with several banks. Usually before our customers go to the bank for loan, we deal with that bank first. We provide them our company profile and financial statements, so that it will make the deal between the bank and customers easier.

Lastly is handling our customer problems. We have received complaints from existing buyers and we will look after their needs attentively. Actually, even if the liability period is expired or not, we will take it upon ourselves to just rectify the defects. Hopefully, our good reputation will be enhanced for future customers. We have a complete record prior to attending to the customers' complaints even if we handed over their keys more than 5 years ago.

7.24 How About Your Internal Process? How Do You Manage It?

To measure company performance annually, we hold one special meeting by the year end and implement internal evaluation based on complaints received throughout the year. This is one of the ISO procedures that we abide by.

I still remember also how well I did when we were facing the economic downturn. Firstly, I held a meeting with my staff and told them that I will increase their pay as a reward to work harder during the recession. At that time, I really took care of the skills on human resource management. For me, manpower is our priority especially during hard times. During that period, my partner told me not to launch any projects, but I went against his idea and launched one development project. Fortunately, the project was quite successful despite it took a longer time to sell out.

7.25 Do You Consider This Business as Risky?

Yes, it is very risky and very dependent on many things such as the economy, environment and so on. During bad economy, banks will not simply approve loans, or maybe if the timing is not right, you have to face a situation where the supply is more than the demand. Both situations require us to be resilient. Let me share with you one story. We used to have a project at Seberang Temerloh Industrial Zone. At that time our country was facing an economic downturn, and prices for raw materials were hiking tremendously. Investors started to sell their shares and no one was willing to invest in any project. But I insisted to continue with our project. We put extra efforts just to make sure that our properties will be sold out, and finally we did it.

7.26 Do You Agree If I Say That Dynaton's Focus Has Diverged from Housing Development to Commercial Development?

Yes, 3 to 4 years back, we focused more on commercial development, but next year we will start with our housing development again. It is like following your destiny, with no specific strategy for us to focus on commercial development only. Our current land banks are suitable for commercial development, so we just proceed with that.

7.27 What Is Your Future Business Diversification?

We will go for hotel as well as food and beverage business, as I see both are more related to what we are doing right now.

7.28 What Is Your Dream in Relation to Your Current Profession Now?

I dream of becoming a mega township developer. Currently, we have a 100-acre township project that started recently. I bought land from my friends, and in the meantime, they are inviting investors to invest in this project.

7.29 Would You Plan to Let Your Children Inherit Your Business?

Yes, either they take over this company or become in-house professionals; both will be fine. We also need professionals like surveyors or architects as I had experienced doing all these things myself. I am the sort of person that pays attention to details; that is why I want everything under my supervision. I always survey other developers' project likes Desa Parkcity and Taman Botanik to gain new knowledge and insights.

7.30 In Your Opinion, What Are the Characteristics to Be a Successful Businessman?

Always strive for the best and be ambitious, so that you can aim high and give your best to achieve them. I am not profit-oriented but rather result-oriented. I will make sure that every single project will be successful and will have credibility. Do not be afraid to take risks. No pain, no gain.

7.31 Can You Give Some Motivations to Newcomers?

I do not consider myself or my company as successful. I always tell my staff to work harder from time to time. Our funding is increasing all the time; hence, our effort should go up proportionally to it. If you think this business is easy, you are wrong. It is better to go for food and beverages or agriculture as both businesses are not so regulated and require small budgets. But if you really want to jump into this challenging business, my advice is to be honest and brave. Do not tell others you can commit to something but are not capable of doing it. This business is risky and stressful, because it involves huge transaction of funds.

Chapter 8
Orchard Vest Sdn Bhd

Keywords Orchard Vest Sdn Bhd • Wong Boon Lang • Banker • Cash flow • Due diligence • Simplicity

Orchard Vest Sdn Bhd is a well-known local property development corporation in Johor with more than 20 years of operation. They have developed eight projects around Batu Pahat and Pasir Gudang. Headed by Mr. Wong Boon Lang, a former banker, the company has been busy preparing for their latest housing projects, comprising Taman Universiti in Parit Raja and Taman Bukit Flora 2, Batu Pahat. Alongside him are three business partners, Chee Kim Hoon, Tony Wang Ee Liang and Ng Siew Poh. Their current development is Taman Universiti which is situated near to Universiti Tun Hussein Onn Malaysia, consisting of terrace houses, shop offices and apartments. Their future development is Taman Bukit Flora 2 that will focus on double-storey terrace houses. Other than housing business, they also involved in commercial development through the construction of Square One Mall, Batu Pahat.

In the business of making affordable homes that delight and satisfy its customer in every way, Orchard Vest Sdn Bhd has built a dedicated reputation for quality. They were created to fill the need in the market place (specifically in Batu Pahat) for spacious, quality built, value-for-money new homes with modern flair. The business success can be strongly attributed to the customer-centric focus that the company takes which is an area of importance and careful attention of the company directors who possess far-sightedness and diligence towards their work. Together with their sister companies, United Vanguard Sdn Bhd and Camellia View Sdn Bhd, they are striving for excellence in reputation, performance and profits.

M. Jaafar et al., *Business Sustainability Model for Malaysian Housing Developers*, DOI 10.1007/978-981-10-5266-8_8

8.1 First and Foremost, How Do You Define Success in the Housing Development Industry?

Success is when your project is sold out and you are able to sustain the business, not only for one project. Besides, your products must be well received by the public and to a certain extent allow you to build a reputation in this field. Another thing is the delivery of good products at competitive prices. In essence, it is not just the monetary definition as it includes every part of development as well as customer satisfaction towards the product.

8.2 Please Share with Us Your Background and the Reasons You Entered This Industry?

My father was a school teacher while my mother is a full-time housewife. I was born and bred here, in Batu Pahat. I am married and have three children. I was a former banker before I entered this housing development industry. I was in the bank for 20 years, i.e. 10 years in Batu Pahat and another 10 years serving the Kluang and Melaka branches, respectively. I have been in this field for almost 25 years. Actually, I did not plan to be a developer. I was involved in this field by chance, being invited by my colleagues to join them as the shareholder of the developer. As I mentioned, I was trained to be a banker until 1990, when the opportunity came and I decided to join this industry. In 1992, I set up this company while, within 1990–1992, I was with other company. As for my educational background, I finished my tertiary education until HSC (now STPM – *Malaysian Higher School Certificate*), and I got a Diploma in Banking. It was during my time in the bank that I took up this diploma course. Obviously, with this credential, it helped me to become a successful developer especially in the aspect of finance because if you want to start housing projects, the most important thing is financing, right? You need money to build the houses and pay all the materials. So, financing is the crucial factor to support your projects. Being a banker, it has helped me to manage finances and plan or foresee the expenses (money) wisely. Also, with such experience, it gives me an edge or advantage to know about good timing for property and subsequently complement my knowledge in this sector. After all, I try to apply the knowledge I gained during my tenure in the bank into my housing business. We may not require bridging loans, but apparently, we still need to deal with banks for end financing towards our purchasers. In fact, most corporate developers go for bridging finance especially those involved in huge projects with 100 units of houses. For my case, we are quite dependent on bank facilities especially during our starting period. The fund will further determine the size of the project. Prior to that, the bank will assess your proposal whether it is marketable or viable. Timing is also important as it can also determine the project size. We obviously have to consider a lot of things before initiating a housing project. Additionally, I am also an active REHDA member, especially for the Johor chapter.

8.3 May We Know a Bit About Your Company?

Essentially, Orchard Vest Sdn Bhd is the parent company, while Camellia View Sdn Bhd is our subsidiary. For the time being, United Vanguard Sdn Bhd is another property arm that concentrates on housing development with me as the major shareholder. Our main projects are mixed development that consist of houses, shoplots and others to cater for the lower income earners. In the future we are planning to venture into property investment and not just build and sell properties but to earn through investments. Our operation is mostly in Batu Pahat but we do also have projects in Johor Bahru. Our project there is known as Flora Heights. At the moment, we only want to become a regional developer and avoid becoming bigger like S P Setia Berhad, Mah Sing Group Berhad or others of the ilk because the property industry really depends on the demand and population of such area. We wish to clearly concentrate in this state only. I can be considered the founder of the entity, but actually it is a shared business (I would say) and not a partnership. It is a company or a corporation.

8.4 Please Tell Us Your Company's Vision and Mission?

We want to build houses that will give benefits to us and others as well. Other than that, we wish to provide affordable homes for the public. So, at this point, we are trying to develop more terrace houses than semi-Ds to cater for the said niche market.

8.5 Do You Find the Housing Business as a Risky One?

I think every business has its own risk. In the business, either you move up or descend, or either you gain money or make a loss. Only this industry is unique as it consumes a large capital that makes it look a lot riskier. But for me, every business is risky. That is why before you commence any housing project, you need to study first. Everything needs to be considered before you purchase the land; also the timing to launch the project should be taken into account. Again, the projects' viability and financing are essential aspects that need to be emphasised before commencing on the project.

8.6 Could You Please Share the Characteristics Needed by Housing Developers to Succeed in This Industry?

First, we have to know what the purchasers' point of view is. Do they look for credible developers? Or do they observe only on quality products? This approach is more apt and will give ideas for us to deliver good products. For me, personally, the quality of the houses should be impeccable with least complaints so that other people will be attracted to buy our products. Here, reputation will certainly be built with good track record. Hence, on-time product delivery and quality of the houses are very important in this industry. Apart from that, in order to attract buyers, we need to leverage on competitive pricing. We have to know our rivals' price before coming up with our own. Obviously, we must keep updated with their strategies as well.

8.7 May You Please Share the Initial Stages of Developing This Company? What Were the Challenges Encountered Throughout Your Journey to Be at This Level?

Again, I would consider this firm as a small one. We started from scratch, small and humble. Perhaps, the genesis is different with other public-listed companies. Our first project is Taman Pantai Baru, which took place around 1995 under Flora Heights Sdn Bhd. We then went for a project named Taman Flora, but we discontinued the firm due to certain shareholders' matter after the completion of Taman Flora. For your record, Flora Heights Sdn Bhd and Orchard Vest Sdn Bhd are two different entities.

8.8 What Is the Company's Marketing Approach? Do You Have Any Special Strategies or Methods That You Can Share with Us?

We have eight staff working under the marketing department. We are using websites and other ICT tools to market our products. Aside from that, we have a sales office and open booth for promotion in shopping malls. We also put on advertisements in brochures and newspapers as well.

8.9 Do You Outsource for Contractors, Surveyors, Architects and Other Related Professionals?

We do outsource for contracting firms and other professionals since we do not have any licence for that purpose. Perhaps, in the future we will have our own, but considering that we are only small in size, we might stick with our initial plan to only outsource the professional services.

8.10 How Do You Acquire Land for Housing Purposes and How Many Land Banks Do You Currently Have?

Most of the time, we purchase the land and apply for subdivisions to get approval from the authority before we can launch the project. Other than that, a joint venture strategy with landowners is another method to secure a piece of land. At certain stages, we need to wait for government compliances and bureaucracy, and this consumes time between the land purchase and the project launching period. Obviously, a lot of challenges can happen within this phase. A lot of money is involved and you may want to seek for bank assistance during this period. In order to purchase land, you need a lot of financing resources, be it your own or from the bank. Again, prior to that, you must look into certain criteria like locational and pricing aspects. It is useful to know whether the project can be sold or not, its marketability, the reasonable pricing and other related affairs for us to sustain the business.

8.11 What Are the Criteria Considered in Deciding the Site Location?

Taking the example of our current project, Taman Universiti, we need to foresee or predict the demand. We believe the population comprising staff and students of Universiti Tun Hussein Onn Malaysia will be our main buyers. The market analysis and procedure needs proper planning and keen observation for us to obtain some ideas about our future projects.

8.12 Are There Any Innovation Being Applied on Your Company or Housing Projects?

We do implement product innovation to attract buyers because I believe it provides competitive edge for us. To me, I always go for simplicity. Basically, we have to adhere to the needs of the homebuyers. Perhaps they are keen into products that are value for money and more practical in design or building structure. I have my own perception, one that it is not necessarily over-designed; at the end, however, the houses are full of leakages, flaws and damages.

8.13 How About the Financial Status of Your Company? The Former REHDA President Said Cash Flow Is Very Vital in the Performance of Developer, What Is Your Opinion on This?

Without doubt, cash flow is very critical for housing development firms. You may sell all your products, but if you do not have cash or adequate funds to pay contractors, for example, I believe your project will be in danger since project payment is usually on progress basis. The construction cost is much more crucial to be settled in order to ensure the project is well completed. Otherwise, your project will slow down. To succeed, you have to consider every aspect until the house is delivered to the buyer.

8.14 As You May Well Be Aware, Economic Cycles Happened Every 10 Years, How Do You Tackle This Situation in Regard to Your Company Performance?

I do not know whether it is valid or not, but having said that, the logical thing to apply is to downsize your construction during bad times. You may also keep the project in slow motion, or you can defer the project until the economy recuperates.

8.15 Does Government Play an Important Role to Help You Succeed in This Housing Industry?

Of course, the government plays a big role in moulding the housing sector. Late approvals by government will lead to obvious predicament. Since this industry is rather regulated by government, it is admittedly getting tougher to do business. In fact, through REHDA, we try our best to mitigate such problems to help the real estate developers because the government rules are always changing. Other things like the role of the OSC (One Stop Centre) and licence applications are somewhat becoming issues nowadays. It does give a slight effect on the project definitely, and it will impede the progress and planning too.

8.16 What Are Current Challenges Perceived as Obstacles for Developers?

The housing industry is extremely regulated these days, especially the recent amendment to the Housing Development Act. While there are some obvious advantages, the responsibility weighs heavily on all the players as any non-compliance can land the developers with a hefty fine and the potential of jail term. The 3% of the total cost of construction of a project less the land cost to be deposited with a licensed bank in lieu of a permit to sell is in itself an added strain on the developers' cash flow position of the project. Other than that, it is also not easy nowadays to procure strategic land at reasonable prices. We therefore need to go suburban to develop but have to brace for the risk of a more uncertain market demand conditions.

8.17 Have You Received Any Awards from Your Housing Business?

I have not received any awards, be it from the government or other non-governmental organisations (NGOs), but I do not consider that to be very important. Anyway, we have only obtained some recognitions and appreciations from NGOs since the company conducts CSR (corporate social responsibility). On top of that, I believed that through our tax deductions, it signifies decent money that we had acquired throughout our business to be returned back to society.

8.18 Any Lessons or Advices for Newcomers Delving into the Housing Industry?

Frankly, the housing industry is not simple. It is not just about building and then selling the house. If you buy the land at higher cost, it will be hard for you to recoup since you will have to put higher sales price that may result in sluggish sales or, worse, failure. Everything needs to be considered, from the time you bought the piece of land until the house delivery. Then, other criterion is the timing to launch and market the product, whether good or bad times. The same goes to financial capability and design whereby every phase has its own importance. Most importantly, cash flow must be monitored and controlled. For newcomers and beginners, they need to be aware and have knowledge about the industry. Hardworking may be the critical aspect to succeed, but it should come together with sound knowledge of the industry especially in handling the projects and problems. Apart from that, good management and careful planning from the starting point are the central features to gain success.

Chapter 9
Taman Jadi (1989) Sdn Bhd

Keywords Taman Jadi (1989) Sdn Bhd • Jadi Group • Heng Tee Theng • Strategic location • Market trend • 'One project, one account' approach

Founded in 1989, Jadi Group is a leading property developer in the southern mainland of Penang, Malaysia. They are based in Nibong Tebal, and their projects cover areas in mainland Penang and northern Perak. The milestone of Jadi stretches back to the late 1980s when they started with a solid foundation with the first 339 units mixed development project in Tanjung Piandang, Perak. Jadi then moved forward to envision and materialise several valuable projects in Nibong Tebal and Sungai Bakap in Penang.

Jadi Group was established by Mr. Heng Tee Theng who conquered many setbacks in life on his road to business success. However, he did not let all these setbacks dampen his spirit. His experience, down-to-earth nature and 'hands-on' approach have sealed his reputation in Nibong Tebal as a trustworthy and prominent land developer and figure in society. Jadi Group is a family-run business offering professional services in real estate development and construction. Their proven track record in relentlessly providing reliable and top quality properties has made them leaders in their own territory. They were created to fill the need in the market place for spacious, quality built, value-for-money new homes with contemporary flair. This focus, backed by nearly 30 years in the building industry and part of Penang's most recognised family homebuilder, has culminated in the huge success story.

Over the years, Jadi has also founded several subsidiaries including Prima Jadi Sdn Bhd, Metro Jadi Development Sdn Bhd and Prima Jadi (PG) Sdn Bhd to cater for multiple growing projects. Jadi has grown sturdily in capacity and experience over the years and have made their presence for a quarter of a century in housing development. Among the notable projects are Taman Pancur Mutiara, Taman Fajar Indah, BM Highland, Jadi Residence and Taman Cempaka. Indeed, from humble beginnings, Mr. Heng Tee Theng has grown his business into a thriving family empire. At the age of 80, he is looking forward to retire from the limelight and hand over his business to his children.

M. Jaafar et al., *Business Sustainability Model for Malaysian Housing Developers*, DOI 10.1007/978-981-10-5266-8_9

9.1 Could You Please Briefly Tell Us About Your Family Background?

I am originally from Sungai Udang, a village near Nibong Tebal town. My father was a fisherman 60 years ago, and at 12 years old I have already followed him to catch fishes. I went to school until Standard 6 only, basically a Chinese school. At that time, I would say that I knew nothing and could not even understand English and Bahasa Melayu.

9.2 How Did You Start Your Own Business?

I started my business since I was 16 years old, selling fishes and prawns. Up till 1976, I managed to export frozen prawns to Australia, Hong Kong and the United States of America. I began with a small-scale enterprise and slowly sought for capital with my partner. At that time, I dealt with people from the Netherlands, Germany and others, but during 1982, the production of prawns deteriorated, and government regulation was getting stricter. In 1983, one of my acquaintances from the United Kingdom sponsored me to go to Bangladesh to see the operation of the prawn industry there. While looking for other opportunities, I found that fishes there were cheaper and so I decided to import fishes from Bangladesh to be sold to our local markets as frozen products. However, in the middle 1980s, several competitors emerged, which prompted me to look to Myanmar as a business alternative. Unfortunately, bureaucratic vagaries hampered me to go further. I decided not to pursue that endeavour and returned back to frozen prawn trading until 1992, when I started this real estate business. Notwithstanding my position as the business owner of this property development firm, I am still involved with marine commodities, but only as a shareholder, while my partner is handling the business. Above all, it deals only with the domestic market.

9.3 How Did You Develop an Interest in the Housing Industry Given that You Had No Prior Related Experience?

Actually, I have been involved in land activities of buying and selling since 1975. At first, I thought this business can really make money, and there was a bank that was supportive and believed on the potential of land development in Malaysia. As a result, I bought a sizeable piece of land in Tanjung Piandang, about 29 acres to erect 339 houses under the name of Taman Piandang Indah. This is my first project and even without experience, I must admit that I had a lot of interest in it. I had to trust my own capabilities to realise the project. Although it was tough at the preliminary stage, I eventually made a success out of it. The main challenges during the initial period stemmed from the numerous parties, as it was so difficult to acquire bank loan and get the layout approval as it took several times to be endorsed.

9.4 What Attributes Do You Need to Succeed in This Industry?

I think it is due to my determination. No matter what happened, I will try my best to strive for success. I failed to secure bank loans initially, but I tried to convince the bank through friends that I worked with during my previous business venture. I believed I could secure the loan given my good business record. At last, I managed to obtain a RM6 million bank loan for my first project. After that, I used my own savings for subsequent projects, as things were getting easier and smoother. Obviously, if I am confronted with predicaments, I will sort them out myself.

If we looked back to my first project, the land grant was for Malays, but I managed to convert it to be sold to Chinese also. That was the only land with an open title, surrounded by other Malay Reserve Lands. I grabbed the opportunity because I foresaw the community had money to buy properties. At the end of the day, all the houses were sold out although sales were slow and gradual.

9.5 With Respect to Your Initial Operation, What Was the Strength of Your Company in Terms of Human Capital and Long-Term Target?

In the beginning, it only involved four persons. My partner and I worked together with one accounts assistant and one marketing staff. At that time, everybody needed to know every single step of the development process. We do not have specific target or vision during the early days as we only wanted to complete our 'first baby'. We then started to be more aggressive by securing land banks in Penang and Perak steadily. Now, we have lands in Bagan Serai, Nibong Tebal, Bukit Mertajam, Tambun and other nearby regions. The largest land bank is located in Tambun, where we plan to develop 1,237 houses under a mixed development proposal comprising terrace houses, condominiums, clubhouses and gated houses loaded with amenities such as swimming pool, landscaped park and others.

9.6 How About the Factor of Competition? What Made Your Products Different with Others?

Every business has competition. They sell their own products and I sell mine. It depends on the buyers. If the buyers like or prefer this location, they will certainly buy it. Location is very essential. Rivals cannot make us successful as we need to be far-sighted on what will happen next. As for my project in Tambun, I am targeting the homebuyers from Penang Island since the Second Bridge of Penang has opened to the public. Our product is beautifully designed and is somewhat cheaper than others. We are ensuring quality without any compromises, so we do expect people will buy our houses.

Despite the competition and escalating cost of land, I believe that with good location, houses can still be sold, and we can still engage in development. On top of that, we need to be flexible and be aware with our environment since this business is fraught with both internal and external risks.

9.7 In the Future, Where Will Your Focus of Development Be? Will It Be in Penang Only?

If there is opportunity, we are keen to spread our business to Kuala Lumpur and Johor. Ultimately, it still depends on the land and price. If it is reasonable and the project is viable, we will definitely go because we have the necessary experience. Perhaps, it would be better to expand since the risks and prerequisites are quite similar in other states, while dealing with authorities and council is deemed to be a normal process.

9.8 What Is the Latest Project of Jadi Group?

It is known as Tambun Royale City, a new township development in Bukit Tambun, Penang, which has something to offer house buyers with different needs. With the tagline *A Township For All*, the new development boasts of a strategic location and facilities suitable for a thriving lifestyle. The township consists of Royale Infinity (three-block condominium), Royale Nova (double- and three-storey terraced houses), Royale Heights (three-storey bungalows and double-storey zero-lot bungalows) and Royale Commercial (double-storey shop offices). It is developed on a freehold site with an estimated gross development value of RM400 million.

9.9 From Your Experience, Which Department in Your Organisation Plays the Most Crucial Part?

All departments are imperative to us since they complement each other. Most importantly, the staff welfare is taken care of. If they performed well, we will offer increment and bonus to incentivise them to work better. In our organisation, we convene centralised meetings every week to monitor their progress. Again, it depends on the suitability and requirements of the department.

9.10 In Terms of Marketing, What Channels Have Been Adopted to Ensure that the Group's Targets Are Well Reached?

We have considered every part of marketing strategies, through Internet marketing, model houses, newspaper advertisements, and booth set-up in hypermarkets and shopping complexes. Occasionally, we also hire agents to sell our products. With regard to the strategy of having booths in shopping malls, we need to analyse the market trend and location of the mall since we need an extremely large crowd to make it all worthwhile. Queensbay Mall in Penang Island would be a very good example and able to attract the right volume and increase public awareness towards our projects.

At the same time, we have our own manager who is accountable for marketing policies. In order to target certain groups of homebuyers, we have to look into each detail of the environment and income level. Other criteria like sufficient infrastructure and accessibility play an important role to assess the viability of the projects. The scenario in Bukit Tambun which is near to Second Bridge of Penang is one where I perceive that the most feasible type of accommodation would be a combination of condominium and landed property. After all, we must follow the current trend and demand.

9.11 Regarding Financial Management, How Do You Control Your Firm's Accounts and Finance to Sustain in This Business?

To have a solid and strong financial ability, we need to meticulously calculate every projected element. For project accounts, it depends on the situation of whether to combine the accounts with the previous projects or create a new account. I personally prefer the 'one project, one account' approach because if they were amalgamated, I am afraid that it would be hard for us to monitor. This procedure also applies to the 'one project, one company' strategy. It entails our subsidiaries to take up new projects for us to easily supervise the Housing Development Account (HDA).

9.12 Do You Implement Outsourcing Strategy or Only Hire Consultants and Professionals to Work with You Under One Roof?

Basically, we employ outsourcing strategy because we only want to be fully responsible for the development aspect. For instance, we appoint a main contractor with competitive quotations and high-quality workmanship. Normally, we open the tender to the public, but sometimes we prefer to work with the existing contractors because of previous experiences with their jobs and the trust already in place. Interestingly, for professionals, once development approval has been obtained, they will invariably contact us for the job.

9.13 How Do You Acquire Your Land Banks? Do You Have Any Malay Reserve Land in Your Collection?

In Penang, this type of land is available but we are not interested to own it. We usually buy open market land purposely for development. We noticed that Malays also bought our houses. Based on current law, we are restricted from buying Malay Reserve Land. Also, if we chance upon agricultural land that has potential, we will attempt to apply for conversion into development land, although this is not a common practice in our organisation.

9.14 On a Separate Note, Do You Consider the Macroeconomic Effects Towards Your Projects?

In terms of economic cycle, it is difficult for us to predict since it relies on the current firm management. I have observed that even huge companies that possess solid financial backups are incapable of surviving during economic crisis. From my experience, during the global recession in 1997, the side consequences impinged on us. What we did was just to slow down production. Yet in spite of that, we still managed to deliver a large project at that time with 300 units of houses. Admittedly, one has to be a little bit of wary, and care must be employed during that period.

9.15 Last but Not Least, Could You Please Give Some Words of Wisdom for Individuals that Intend to Dabble in the Housing Development Industry?

Actually, I do not think that I am qualified to talk about this. They have to start small-scale because this industry is getting difficult to conduct and handle. During my time, I regard myself as the luckier one to be able to succeed and reap the fruitful efforts. They also have to be brave and need to know the intricacies of this business before delving further. They must consider the role of the government; sometimes the governments are in our favour while, most of the time, they are against us to protect the benefits of the homebuyers. With fluctuating rules and changes made by the government, the industry is fraught with high risks and challenges. Nonetheless, we still have to face it by thinking positively and daring even to fail.

Chapter 10
PDG Property Sdn Bhd

Keywords PDG Property Sdn Bhd • Gwee Yew Kiat • Entrepreneurial learning • Firm reputation • Site identification • Strategic planning

PDG Property Sdn Bhd is an award-winning company which has been predominantly engaged in the business of residential and commercial property development in Melaka since its incorporation in 1994. This company is a leading property developer in Melaka, standing at the forefront of building quality by concentrating on affordable housing in strategic locations that caters to every stratum of society. With their vision of bringing property to life, PDG Property has successfully completed more than 30 projects, with many ongoing projects and many more being planned, spreading throughout Melaka and Johor. Besides that, it is also their aspiration to always improve on project development and provide more quality homes and services to all purchasers.

The current Managing Director, Datuk Gwee Yew Kiat, has been in the property business for more than 20 years. He is a member of the National Council of REHDA Malaysia and also an Immediate Past Branch Chairman of REHDA Melaka. It is also noteworthy to add that Datuk Gwee Yew Kiat has played a prominent role in developing MCL (Malay Customary Lands) which are in abundance in Melaka. These lands which are mainly located in the city centre as well as outskirts of Melaka have placed PDG as one of the major contributors in housing needs for the Malays in and around Melaka. The regular engagement and brainstorming sessions carried out by Datuk Gwee with his consultants and counterparts are commendable as they have produced many creative designs, practical use of space in housing and also to have the best use of land in planning for new developments.

A far-sighted developer with social aspirations for building homes for the needy, Datuk Gwee has placed himself in the heart of the state government of Melaka. This has resulted in him being awarded two abandoned projects in Melaka to be redeveloped. With his vast experience in housing development, it took him less than 2 years to turn around these two projects from being labelled 'abandoned' to sought-after projects. The abandoned projects are Taman Perindustrian Ayer Keroh encompassing a total area of 69.49 hectares with a total of 1056 units and Bukit Baru Business

M. Jaafar et al., *Business Sustainability Model for Malaysian Housing Developers*, DOI 10.1007/978-981-10-5266-8_10

Centre with an area of 6.04 hectares. Taman Perindustrian Ayer Keroh was approved for industrial developments way back in 1992 and left abandoned a couple years later. With his foresight, Datuk Gwee redesigned the components and injected new ideas and came up with mixed development comprising high-end terrace houses, semi-detached bungalows and also factories and industrial lands. This project will reap a total gross development value (GDV) of RM500 million. Meanwhile, Bukit Baru Business Centre which was originally approved for condominium was also transformed into mixed development. The new project comprises 121 units of 2- and 3-storey shop offices in Phase I and 3 blocks of condominiums totalling 400 units in Phase II. The Phase I has been completed, while the condominium block is still in approval stage. This development has a total gross development value (GDV) of RM200 million.

10.1 Could You Please Describe Your Educational Background?

Actually, I am not a graduate because I am just a school leaver. I then sat an examination to become an insurance agent before moving to the property development business. Personally, I will not stop learning since I love to learn new things.

10.2 How Did You Start Your Career in the Property Development Business?

Prior to starting the business, you should learn the history of the industry. You need to understand issues related to the economic crisis that struck Malaysian in 1997, when big companies such as Sunway Berhad and Country Heights Holdings Berhad suffered from their investment in land banks. It took almost 10 years for them to recover. I should thank our previous Prime Minister, Tun Dr. Mahathir Mohamad, for managing the crisis very well. If I am not mistaken, it took a shorter time for the crisis to recover compared to the economic downturn that occurred during 1985/1986.

I started my working experience as an insurance agent and underwriter during the age of 20s. I did general insurance, especially motor insurance, and I worked on it very aggressively, to the extent that my business dealt with millions in premium every month. I would consider that I had managed to control the market at that time as most of my agents were motorcycle shops and used-car dealers. For your information, previously the Central Bank of Malaysia (BNM) allowed the role of underwriters since they were appointed by insurance company and acted as chief agents. The insurance company would give commissions to the underwriters, and they will subsequently distribute the commissions to the sub-agents. As underwriters, we collected premiums with a tight policy and settled claims accordingly.

Through my experience as an insurance agent, it allowed me to explore myself as well as the opportunity to meet people. My father passed away when I was 17 years old, and I had to struggle during the initial stage. Due to our hard work and blessings from God, we managed to secure a big volume of business. I was lucky because one of my networks in insurance was a used-car dealer that used to be a contractor, and he offered me to join him to establish a contracting firm. But, we found that construction business cannot make a lot of money. Hence, we switched to housing development. I had a friend who was a renowned footballer, Datuk Soh Chin Aun, who wanted me to join him as a business partner. From his connection, he bought a piece of land, about 5 or 6 acres, for us to concertedly develop Taman Seri Jati with 85 units of houses as our first housing project.

I started my first project when I was in my late 20s. No banks were interested to fund our project. In the end, we got an approval from Hong Kong Bank as they were prepared to give RM1 million for that first project. That amount however was just enough to buy the land only and again our project was delayed. Luckily, a friend from a reputable bank helped me with the bank loan. Eventually, the project was a success and I never look back ever since. From there onwards, we are careful about buying land banks as we are sceptical with abandoned projects. In this case, I really agree with the regulation to make it compulsory for every project to have Housing Development Account (HDA) in order to ensure that the particular money will not be misused.

10.3 So, Are You Telling Us that Those Who Are Tied Up with Loans Will Be Facing Problems?

Not necessarily, because it depends on timing too. Land price is hiking up by 200–300% lately. So, when we bought this land 10–15 years ago, we were very lucky as we are going to make a profit. But if you buy land at its highest price or when the market is overbuilt, you will face problems to sustain. In order to avoid this, developers nowadays change their business strategy to launch their projects by phases. Previously, they launch all available units, and as a result, they faced problems to sell all units within a stipulated period. Taking YTL as an example, they started by launching their 30 units of semi-D houses in Kuala Lumpur, while S P Setia, in Johor, started by launching the first 100 units of their project.

10.4 What Are Your Company's Strengths?

Our strength lies on our strategic planning. Since the property business is our priority, we acquire as many land banks as we can. Our focus is on small pieces of land because it is easier for our company to develop it immediately after the acquisition process. The demand for houses is still there even though the population in Melaka is not that large. By focusing on small-scale development, we can ensure that our units will be easily sold.

We did some studies and found that developing a housing project near to the existing residential area is a good move as young generations prefer to own a house near to their parents especially for non-Bumiputera. Another factor is the development area. We should consider the accessibility and traffic problem, especially on working days.

10.5 May You Please Share with Us Any Innovative Efforts that Have Been Applied by Your Firm?

While a three-room house is common nowadays, we took the initiative to build four-room affordable houses to be offered to the people of Melaka. We know from our analysis that the saleable price is below RM200,000 or RM250,000 for semi-detached units. Hence, our product is in demand for a big family with middle-level income. Besides that, in total, developers in Melaka only build about 4000 units per annum, while demand is higher than that, with about 7000 units per annum. Therefore, this is a chance for development firms to build more houses with affordable price.

10.6 Do You Have Any Target Per Year that You Want to Accomplish?

We consider ourselves a small developer, with target sales value of RM50 million per year. However, with these current and upcoming projects, we are targeting RM100 million of sales value for each project. For the last few years, our business is encouraging and doing well though.

10.7 Is There Any Person that You Consider Him/Her as a Mentor? Have You Received Any Awards Along Your Journey in This Field?

I am doing small-scale projects, which basically anybody can do, so I do not need any mentor to advise me or direct me. My companies, Johan Pertama Sdn Bhd and Naga Jaguh Development Sdn Bhd, have received two awards, the Golden Bull Award by Nanyang Siang Pau in 2004. They were also listed among the top 100 SMEs in Malaysia. As for PDG Property Sdn Bhd, it was awarded Golden Bull Award in 2012 and Golden Eagle Award in 2013. Indeed, it is a normal practice for any development firm to have a number of small-scale firms under one roof or parent company. This is one way for us to segregate potential risks and prudent tax planning. Furthermore, it is a bit difficult to apply loans for a big volume project. By dividing it into several companies, banks will have no problems to release the loan as bank foresees different companies as different entities. Sometimes, there are projects that have different partners too. To make it easy, we form a new company and give the new partners certain percentage of shares in the company.

10.8 How Do You Secure Your Land Banks?

We actually buy the lands. Acquiring land by joint venture will create more problems, as people tend to change their mind when dealing with money. I mean they are prone to betray us. I used to have several experience in dealing with these cases.

10.9 What Motivates You to Venture into This Housing Development? Will It Be for Profit or Interest?

Profit of course is one of them. For me, housing development is the best business I have ever been involved in. Actually, when I was young, I did many businesses such as restaurant, boutique and other things, but it cannot be compared with development business as its income is higher than others. The profit is very large and the volume is there with less cost. Previously, I did a 50-million business with only four staff, which is very cost-effective. I do not see any other business that can do this. Construction requires a lot of staff, which is why I prefer to outsource or subcontract to other contracting firms and concentrate on development only. My thinking is as long as your house can be sold, then you can make some profits.

10.10 How Do You Assure the Quality of Every Project?

Firstly, we get the best design for our houses. Our duty then is to ensure that when our buyers want to resell the house, they will get a better price. Currently, fellow developers are following our design, especially our cluster house design, which is indeed four units attached together, so that every unit is a corner unit. Security-wise, this design is better than a semi-detached unit. My designs are mostly practical ones that benefit the residents. I believe that most of the purchasers are first-time home-buyers. Hence, we try to explain to them the rationale of our design, which will include better features, with four rooms, i.e. all are square in shape and have wet and dry kitchens to suit Malaysian style of cooking.

In order to assure superior quality, I also refer to other countries to get better designs, such as Australia, with some modifications to fit in Malaysian culture. Our architects always put themselves in the customers' position, designing the best model to fulfil the customers' need. To be frank, even though our prices are a bit expensive, there are still many customers coming over to buy our products.

10.11 On a Related Note, How Do You Market Your Projects?

Normally, we will build a show house first and then proceed with the promotions. A conventional way sees us opening a booth for promotion at shopping complexes. As communication technology becomes easier, we do online promotion to tackle younger generations as our target group. Our target is local people since our projects are not huge.

Non-Bumiputeras will immediately take up our units once advertised, even to the extent of giving their names in advance for our future projects. For our projects in Melaka, we always take advantage from the nearby newly launched projects by other developers. It is due to the real situation of residential development projects in Melaka which are usually small or medium in size since it cannot cater for the current demand. Ironically, in Johor, the population is large enough for the developers to consider developing a township; honestly, we are currently running out of land banks, and only recently we bought land in the range of 20–30 acres at Kluang, which is not large size-wise, but we need to be prudent in this matter. If the economy is still slow, we may need to defer the project. Had we bought a bigger size of land, we could be in trouble as we will have to pay on a higher loan especially during economic downturn.

In light of this, we are very conscious with the Greek and Europe economies. If the situation continues to worsen, it will affect our economy indirectly. Less imports from them means less income for our country, and it will instigate recession of the economy. It will affect us in a way that people cannot afford to own a house since the price for property is constantly increasing.

10.12 Since You Are Now in the Midst of Venturing into the Johor Market, Provided You Do Business Through Outsourcing, How Do You Manage Your Business at that Particular New Market While All Your Subcontractors Are Based in Melaka?

Actually, this development industry is very simple and straightforward. Once you bought a land, you may start finding your contractors, consultants and local engineers. It is not a big problem as we have more than enough professionals in this particular industry. When we do projects in Johor, we employ local people. The only precautionary step that must be taken is to regularly go to site and do checking and researching on the needs of the local people. In a nutshell, to make things possible, do mingle around to know them better.

10.13 We Would Say that Johor Is Progressing Very Swiftly Especially in Iskandar Malaysia: Do You Consider Any Threats from Local Competitors?

Nowadays, buyers always look up at big developers such as S P Setia and IJM, and people will think that small-scale developers like us cannot compete. They are totally wrong, as we normally do not compete with them. Firms like us focus more on development on the second layer of the town or suburb projects, while the big companies are looking forward to develop townships; hence, we aim on developing the nearby places with 20–30 acres of land. Taking Kluang as an example, the biggest developer there is BCB Berhad and for sure we cannot compete with them. But luckily, they are now moving their business to Kuala Lumpur, and this has certainly opened up opportunities for us to do more business in Kluang.

10.14 How Do You Create Your Firm Reputation or Branding?

We create our reputation through quality and design. We are honoured when we know that our customers had introduced us to their friends and family. It shows that they believe in our quality of work. That is our strength. In this business, I also noticed and learnt that you cannot do mass products. Suffice to say, you need to be brand conscious. If homebuyers feel confident with us, they will promote us via word-of-mouth recommendation, for example, '…this developer gives your houses in time, no bad records and good design as well'. If we are already reputable, buyers will put their trust on us and it will be easier to sell our houses in the future. It is because of our good name in Melaka that the Malays are starting to buy houses from us. I believe it is due to our good workmanship and quality.

10.15 If You Do Not Mind, How Much Profits Do You Make for Every Project?

Not that much, around 20–30%. Some of my projects even recorded less than 20% profit margin especially for projects in the Malay Customary Land (MCL). At times, we also build affordable houses, and the recorded profit will be very much less, about 15%, but the volume is much greater and easier to sell. After all, this type of project needs a lot of capital, but with our good reputation, it would not be a problem to get bank loans.

10.16 In Terms of Government Role, Does It Contribute to the Success of Housing Developers?

Actually, I would say 'yes'. In Melaka, we could not build much due to its tight quota. However, the government helps much nowadays. So far, we seldom have abandoned projects in Melaka. Perhaps, developers here tend to think seriously first by doing proper analysis about the project. The only one I want to address here is about the government loan process or approval. It usually takes a longer time, and in some cases, the loan was still not approved even when the project is completed. That is why there is the need for developers to have funding backups because it is very imperative as we need to use our money first since we can collect very little deposits from purchasers.

10.17 We Would Like to Know About Your Financial Management Strategy? Does It Contribute to Your Success?

As an experienced housing developer, we can handle our cash flow very well. Like normal projects, the non-Bumiputera components can subsidise the Bumiputera quotas but not for MCL projects. We have to depend on bank loans as we cannot claim monies immediately from project sales. As for me, financial strength is important; if not, you would not make it in this industry.

10.18 What Are the Business Principles that You Hold to Ensure the Sustainability of Your Business?

I do not really know actually, because no specific principles have been rendered. And somehow, it could be associated with 'luck'. Perhaps, maybe, our forecast on the location is very good. What we have foreseen seemed very accurate with good analysis on the project such as the affordability of buyers as well as strategic locations. We must know the prevailing market trends and people's preferences. Thus, our projects will be more saleable. Having said that, we appear to be targeting the right groups and their needs, and in reality, we must be attentive with the current choice and requirements of the people. That is why you see our houses are marketable and become bestsellers because we build affordable houses that suit with people's needs and we give the best quality. Personally, I reckon that to measure your achievement is by looking at the take-up rates in selling the housing units. If units are taken up within 1 year, it can be considered successful, but not if longer than 3 years.

10.19 How Many Local Developers Compete with You Here, in Melaka Especially?

Not so many since we are protected and due to high quota for Bumiputeras, outside developers are wary about coming here. Here, we have other fellow developers like Teobros Development Sdn Bhd and PB Realty Sdn Bhd, while GJH Construction Sdn Bhd is more focusing on high-end projects, i.e. bungalows and others of that ilk. We are developers of terrace houses and not so many competitors are in this class of projects. Currently, we are developing cluster houses which put us as the sole developer to produce this type of houses. Eventually, we still welcome healthy competitors to make this industry a better one.

10.18 What Are the Business Principles that You Hold to Ensure the Sustainability of Your Business?

[illegible]

Chapter 11
Fairview Development Sdn Bhd

Keywords Fairview Development Sdn Bhd • Muztaza Mohamad • REHDA Malaysia • Diversification • Working culture • Business mindset

Fairview Development Sdn Bhd was incorporated in 1985, when two enterprising professionals agreed to pool their resources together to undertake and promote business activities in the field of project developments, project management and consultancy services, in line with their expertise and experience. Over the years, the group has undertaken over 2300 units of various types of development, of which 2000 units have already been completed, delivered and occupied by purchasers. The balance of the units is in various stages of construction.

Fairview philosophy is "We Believe in People". In order to succeed, Fairview believes in putting people first. Human development is important both for preparing for the good or worst environment. However, in order for the people to be successful, he/she must put the organisational goals as a guide. Achievement of personal goals must be through the organisational goals (win-win formula). There are a lot of opportunities available for those who want to succeed within the group. In order to become an established development company, Fairview has set the vision "to develop successful people within and outside the organisation to drive the group business". Fairview's vision is to create a company specialising in its business sectors with exceptional quality products and services and at the same time gives importance in human development as well as product development. With the mission statement "to deliver products and services that are a class above others", Fairview will strive to be inventive and innovative by maximising the full potential of technology. They envision to be able to stand out by reason of effectiveness, ability to anticipate and agility in delivery as they believe in partnership, worldwide.

Some of the successful projects that made Fairview well known to the customers are Taman Sri Tanjung, Taman Bangi Indah, Taman Semenyih Jaya, Menara Bakti and Kelana Parkview. An interview with its Founder, Datuk Muztaza Mohamad has revealed few success factors in business growth that may be useful for future undertakings.

M. Jaafar et al., *Business Sustainability Model for Malaysian Housing Developers*, DOI 10.1007/978-981-10-5266-8_11

11.1 Datuk, What Is Your Background and How Did You Jump into This Field?

I started by joining Shah Alam Properties, a property development firm, as an accountant. Prior to that, I was with timber company. At the end of 1983, I started thinking of opening a business but only on 1985 I got a chance to realise my dream. There is no easy way to success. I had founded a few companies with different partners before this but failed. I ended up with my very own company, Fairview.

11.2 Has Anybody in Your Family Involved in This Business?

No, I am the only in my family. My late grandfather worked as a sailor in Melaka, while my father worked as a clerk, also in Melaka. So, I did not inherit any businesses from anybody. But my father prioritised a lot on education. Thank God, many of my siblings are now businessmen, but most of them are professionals like accountant, lawyer and surveyor.

11.3 How Did You Constitute This Company?

One of the biggest partners was the late Dato' Baharudin Osman. He injected fund to my company but more importantly was his willingness to assist in forming this company. He helped out in terms of project management, dealing with government requirement, joint venture and so on. He was the Managing Director of Sime UEP Cum the Director for UDA Properties. His strength lies on human aspect. I still remember when we started, with more stuff to do, we really need his motivation to steer us along his network. He gave us advice as well as guidance on how to run a company. After everything is going well, I have to take the full burden to handle this company. He can be considered as a mentor to me. After all, at the end of the day, if you want to succeed, 99% depends on your own effort.

11.4 Have You Involved in Professional Bodies Other than REHDA?

Previously, I was the President for CIMA and EXCO Member for the National Consumers under MIA Group. I was also an Academic Advisor for School of Accountancy of Universiti Teknologi MARA (UiTM). Besides, I was the Chairman for properties division under *Dewan Perniagaan Melayu Malaysia* (Malay Chamber of Commerce Malaysia) even though I was just active for 3 years only. I prefer to get involved in REHDA rather than DPMM. Through this involvement, I manage to apply corporate social responsibility (CSR) to give back to society.

11.5 Have You Receive Any Reward from Authority or Government?

I received state honours from the Selangor government as well as the Datukship award in 2012 and a certification from CIMA, that is all. But for me, by the end of the day, it is your contributions to the people around you and the society that apply.

11.6 How Did You Get Started in This Business?

I started with no capital since I just worked as a government servant with insufficient savings to start up new business. So, we set up another company to back up our finances. Those companies involve in project management and manufacturing sector. Then, we did a joint venture with the government to raise our capital, so that we can roll the money.

11.7 How About the Diversifications that You Have Initiated?

No, the diversifications were generated simultaneously, and partners for every company are different. My share was just RM10,000 at that time. Some of the partners have already migrated, so we have to restructure our diversified businesses. We do not have any formula on what business to venture as we just grab whatever opportunity thrown at us. It really challenged me, but I like it because it is very mind-refreshing for me. People in countries like Singapore, Taiwan and Hong Kong inspire me; how can they be a developed nation and giant in economy whereas they do not have natural resources as much as Malaysia? Perhaps, you have to transform yourselves into "illegal immigrants'" mentality, work hard and become independent. We have to change our mindset with no privileges thinking to survive in this world by developing business acumen. You must have strong foundation and again, it should start with the underground. It is imperative to highlight this foundation which I reckoned the schools or educational institutions here has failed to produce solid characters of students. Sustainability is linked with the background. We believe that the foundation must be there.

11.8 What Trigger You to Diversify the Business?

It is because I anticipated that this construction industry is uncertain, while my packaging factory seems to be more stable. So, I think diversification is the best strategic business plan for me. I try to make all my businesses profitable by supporting each other as they are mutually complemented in order to maintain company's income.

11.9 What Drive You to Sustain in This Business Despite All Challenges?

We asked Allah the Almighty and believe in Him especially when you want to start up a new business. I think it is the same with other religions, especially Chinese. Their faiths are strong and they believe if they do their best, their God will help them. I am confident that Allah will help us if we do our best and work hard striving for excellence. This is one of the philosophical thinking that Malay should observe as they always give up facing the obstacles.

11.10 How Did Your Knowledge in Accounting Help You in This Business?

Since I am a chartered management accountant and also the President of CIMA, I would say that I am capable of managing the resources. We know how to optimise resources like staff and machine. How to utilise human capital is by giving them training, encouraging them to give opinions and promoting positive culture. In fact, our company vision is to develop people that have values in their lives. The same goes to the university branding where it combines gradual steps to achieve good name and reputation. That is where we are heading right now.

11.11 Is Money Your Motivation to Get Involved in This Business?

Not at all because I have my own philosophy. We want to make people happy by providing comfortable homes for their future investment.

11.12 May You Please Elaborate on Your First Project?

Around 1980s, we had one joint venture project with MBPJ (Petaling Jaya City Council) at Section 14, Petaling Jaya, and then another one at Batu 6, Bangi. At that time, although it was a low-cost project, the government allowed us to collect RM10,000 from each buyer. With 200 units, you can assemble RM2 million already. Not to misuse it, we put the fund into the project. Slowly after that project, we build up company by planning properly and forecasting accurately. It is important for us to have annual projection especially in terms of finance. In addition, setting right goals and following the guidelines are also the key criteria. To monitor the progress and target, we had monthly and even weekly meeting to recheck our progress.

11.13 Can You Share Any Challenges that You Ever Faced?

Challenges are always there but we must learn from the predicament. Do not look at the challenges but look at the satisfaction and enjoy after you overcome the challenges. During economic downturn in 2008, we were facing tough challenges. Due to fund problem, we had to downsize our project, from developing double-storey to single-storey houses. We were transparent to our staff; they knew my savings and also company profits. We treated them as share partners; that is why they understand when we cut the salary of those who earned above RM3000 in our company. We also share the problem with staff, so that they felt that they are belonged to the company. With this method, we can alleviate our problems as well.

11.14 What Is Fairview's Strength?

Our strength is we are very focus to deliver the best products. Since the realm is very localised, we already know the local authorities and the inside out of this area. Knowledge is power; that is why you must master the knowledge on what you are doing. We also acknowledge those who contribute ideas as our strategy for us to be sustainable.

11.15 Why Do You Think People Will Buy Fairview's Houses?

We are well known and customer-friendly. Our after-sale service is good and we always entertain their problems attentively. Our price tag also is competitive and affordable. Our customers keep talking about our products, and that is why people know us from word-of-mouth marketing. Unfortunately, now we do not have much land banks. We have to find it, regardless how small the land bank is, or at least we will go for joint venture.

Currently, we are depending on our rental business and also our factory. We also have car park operation which we can share the overheads too. Later on, we will come up with our development projects; it is currently under pre-development period.

11.16 Other Development Companies Started Their Business with Sufficient Capital and Land Banks, While Your Company Is Otherwise. Could You Please Say Something on This?

We are not a landowner; we are only a developer. Yet, we do not have much capital and it was difficult initially. You must apply the method of interdependency with others. So, we have to sit with the landowners and convince them to buy our services.

11.17 Just Now You Mentioned that You Treat Your Staff Like Your Partners. Is There Anything You Do to Value Your Staff?

We always create opportunities to our staff. For example, we gave chance to our lorry drivers to have their own lorry. To those who worked hard and performed well, we also offer them some shares when we opened new companies. With this profit-sharing concept, it engenders sense of loyalty. We do this because we need these strategic persons to keep advancing. In business, we have to redefine what we want in this field to be sustainable. We must be very practical in each aspect to ensure viability.

11.18 Why Do You Think Malay Is Not Successful in This Business?

It is nothing to do with risk, but more to our culture actually. Our parents yesteryears did not prefer us to work in private sector or run our own business. They would prefer us to work in government sector. My view is different. We should teach our children to do business and encourage them. But do not let them to be too dependent on loan to make businesses. For me, I always pay everything on time; thus we do not have any problem related to supply chain.

11.19 How About Innovation? Do You Implement One?

We can innovate at all aspects of business, including the way of thinking. We can work through outsourcing or interdependence of people. It depends on how do you manage the environment and resources as well as how do you handle people around you. We should think outside the box, manage resources effectively and handle business wisely.

I urge all businessmen to follow Chinese style of business and integrate it with our culture in accordance with Islamic teaching. Malays nowadays still depend on the government and cannot stand by themselves. That is why I prefer to work with Chinese since they are more proactive in this corporate field, transparent as well as systematic in doing work.

11.20 As a Supreme Committee Member of REHDA, How Do You See Government Influence on This Industry?

Now, I act as the Secretary General of REHDA. We frequently have meetings with the government and its agencies. So far, the government is being good to us, approachable and sensitive to our demands. But there are still rooms for improvement as the welfare of property developers in Malaysia seems to be neglected unreservedly. After all, we need to educate buyers, so that they may understand our situation and current procedures or regulations. Do not let the buyers control ourselves or else the public will become ignorant.

11.21 How Do You See Today's Economic Situation Could Affect the Survivability of Housing Development?

It totally depends on them, nothing to do much with economic situation. It relies heavily on how you organise your company, how you plan your strategic planning and how your interest affect your decision-making. This industry also hinges upon demand and supply. More or less, this business is depending on the level of thinking portrayed by the owners.

11.22 How Do Competitors Here Give Impact to Fairview?

I do not think rivals can influence our business growth. For me, the most important is acquiring as much land banks as you can and how can you get into your own market.

11.23 Any Lessons Learnt that You Wish to Share with the Newcomers?

Life is much tougher than before; same goes to this industry because of the laws. Cost of living is very high nowadays and the housing market has matured. Price of houses and land is also very high, but the demands still exist although the scenario is totally different. Ironically, newcomers nowadays have their own strengths because basically they already have experiences in other business or field like restaurant, timber or mining. I think this situation has given advantages to them since they may be having money to invest in this capital-intensive industry. Totally different from my situation before, without money in hand, projects and sufficient staff, it was really hard to survive. Believing in Allah, hardworking, focus and self-centred towards your work are the key recipes for you to succeed. Believe me, nothing is impossible.

11.24 Any Other Business Philosophies that You Want to Share with Us?

We would love to develop the potential of people because we believe that every person has potential to be successful, just the matter of how to polish them. We have to train them to be more resilient in order to strive and progress forward. Suffice to say, they must have strong foundation and be consistent towards excellence.

Chapter 12
Sri Anika Development Sdn Bhd

Keywords Sri Anika Development Sdn Bhd • Leong Kin Kun • Public relation • Business mentor • Positive traits • Marketing research

Sri Anika Group was established in 1995, with core businesses in property development and investment, manufacturing, timber logging, international trade, foreign manpower outsourcing and employment agency. Presently led by its group Managing Director named Dato' Leong Kin Kun, Sri Anika is one of reputable Perak-based companies who has developed a number of housing projects located in Ipoh, Kuala Kangsar, Batu Gajah, Sungai Siput, Chemor and Taiping. Prior to 2008, the development activities were handled by his late father, Mr. Leong Kam Cheow. Among its completed projects include Taman Uda Sofiah, Taman Villa Gahara, Taman Seri Delima, Jalan Taiping Business Centre and Jalan Dato' Sagor Business Square (in Kuala Kangsar), Taman Changkat, Taman Seri Wangsa and Taman Saujana Indah in Batu Gajah. Driven by the heightened spirit of success, Dato' Leong Kin Kun prevails his experience and vision in further developing his business.

12.1 Before We Dwell on Further, Please Provide Your Own Views on the Concept of Success in This Housing Business

Success in housing business entails a combination of positive aptitude, in-depth industrial knowledge, vast experience, strong vision, financial strength and insight from past lessons. A successful developer must be capable of building up an end-to-end talented professional team from the initial planning phase, followed by approval, financing, construction and eventually project marketing phase in order to strive for financial and operational excellence while maintaining customers' satisfaction level.

M. Jaafar et al., *Business Sustainability Model for Malaysian Housing Developers*, DOI 10.1007/978-981-10-5266-8_12

12.2 What Is Your Highest Level of Education? Does It Help or Contribute Towards Your Success in Business?

I graduated with a Bachelor's Degree in Marketing (UK) and found that it is beneficial for my business. The knowledge gained from my qualification has enabled me to build up my business more efficiently and prepare me for the exciting opportunities ahead. Business-wise, I have been participating in learning and development programmes for continuous self-improvement by attending trade fairs, seminars and workshops. By having a comprehensive knowledge of core concepts, updated technology and technical skill sets as well as industry best practices, I will be able to strategise and develop effective marketing practices for my business.

12.3 What Do You Think About the Contribution of Previous Related Experience Towards Business Success? Do You Mind to Tell Us Your Experience Before Dabbling in This Industry?

Prior related experience is definitely a key ingredient of business success. For your information, before I enter the property industry, I have started my first business together with my younger brother, Leong Kin Kee, in 1997. The nature of that business is employment agency. I believe that the housing industry has bright and excellent prospects ahead. Hence, I have diversified my existing business into housing industry by initially doing property investment in 2008. I started to be a full-fledged housing developer after the demise of my late father in 2009. I have been entrusted to carry on with a few housing projects which were inherited from my late father. While my father was still around, I used to assist him in dealing with the government agencies, handling documentation, etc. With those experiences that have made me responsible to carry on my father's business, it has also led me to set up few more companies specialising in housing business on my own initiative.

On a different affair, our customer basis in employment agency is considerably strong since we deal with umpteen sectors such as manufacturing, construction, agricultural, plantation and service industry. As a result, we have strong customers networking in hand whereby some of them are active property developers and contractors. Thus, during some stages, we are able to learn the industry problems from our customers' experience. For example, once our clients apply for additional foreign labours from our agency, they need to fill in all the related documents. From that particular process, we manage to study the whole procedures of various industries. Sometimes, due to tedious and complex practices, we have to perform extra works by studying and preparing additional profile for our customers. Those are the opportunities for us to capture and gain supplementary experience of different type of businesses. I learnt a lot from them specifically in terms of documentation and administration. In fact, we got to know the suppliers and officers that are normally in charge of certain matters. This is indeed a useful knowledge to us.

12.4 Do You Have Any Personal Target in This Business?

I have set a target with comprehensive business strategies to propel my company as one of the most popular housing developers in the state of Perak. So far, I have completed four projects with hundreds of property units (both commercial and residential). Furthermore, my company is currently developing a total of 345 units of houses in Ipoh, Batu Gajah and Kuala Kangsar.

We have also set up a joint partnership with the state government. At the moment, we have several residential projects encompassing 100 units of houses in Batu Gajah, Kuala Kangsar and Ipoh. They are still in the preliminary stage; most of them are in the middle of earthwork activities after receiving the approval from local authority. In terms of land bank, we have few other parcels in Taiping and Sungai Siput for our future developments. With about 600 units of houses to be erected soon, we need ample time to digest such projects.

Due to our moderate firm capacity, we take about 2 years to complete one housing project. For us, the 2-year period is measured from the beginning of construction works to the project completion. It also includes continuous analysis of current trend, marketing activities and other buyers' affairs. Indeed, we intend to move faster than 2 years, but since we operate in developing areas like Batu Gajah, the market pattern is different with developed regions such as Klang Valley and Johor Bahru. The movement of homebuyers is relatively slow here. Nevertheless, with intensive marketing, we still manage to sell out our properties.

12.5 Plenty of Outstanding Entrepreneurs Have Admired Few Notable Persons as Their Mentors. How About You?

I was fortunate to have my late father, Mr. Leong Kam Cheow, as my best business mentor. My father taught me how to run a successful business when I graduated from university. He had enormous achievements in all businesses that he pursued as he was a man with full determination to succeed. Above and beyond, I personally admire one of the successful Malaysian entrepreneurs. He is from Perak too and his name is Tan Sri Jeffrey Cheah, the Founder and Chairman of Sunway Berhad. I find that he is an ambitious leader with dynamic initiatives.

12.6 Business in Housing Is Rather Unique as It Consumes Large Capital with Late Return and the Profit Is Entirely Based on Sales; What Are the Characteristics Shall the Housing Entrepreneur Possess in Order to Succeed?

I personally think that a successful housing developer must be imbued with a lot of positive traits including excellent communication and networking skills, highly innovative, strong passion at work plus a well-planned project that is favourable to consumers (in terms of pricing, location, promotional activities, building design and quality). In my opinion, a successful housing developer must possess good conscience and be self-disciplined. Having said that, I personally think that a legitimate and integrated-based approach shall be applied as a key to sustain the business success.

12.7 Do Strategic Management and Planning Give Impacts Towards Business Success?

Continuous planning and control activities on top of ongoing excellent management support are the major success factors in building a successful business. For instance, as property developers, we need to employ complete assessment of latest market trends on consumers' preferences, price range, location, economic condition, costing, financing and so on. Also, before we determine the project location, we will conduct market segmentation research on demographic variables such as age, occupation, household size, geographical analysis as well as consumers' behavioural study. These factors will be taken into account before decision is made whether to progress with the project or not. In this case, intelligence that comes from experience and knowledge may facilitate us in producing business success.

12.8 At Present, How Many Staff Do You Have?

I have about 50 employees at the moment. They are designated to work on management activities, marketing programmes and construction work, respectively.

12.9 Does Your Company Utilise Any Marketing Strategies? Please Share Any Specific Promotional Tools that Have Been Implemented

Our marketing strategy is drawn from regular and comprehensive marketing research. It focuses on delivering suitable products based on consumers' demand to greater profit margin. Generally, our marketing team always plans for a series of promotional activities which include e-marketing through various applications, distribution of leaflets, mass media advertising in printed publications and online marketing via social media network. Various ICT inventions such as smart phone applications, computers and software applications are commonly used in our businesses to maximise work productivity and aid in sales promotional activities. In our increasingly connected society, we have also built our presence in social media network as a powerful way to market our businesses to customers. Apart from that, I must admit that public relations (PR) is also one of the key fundamentals towards business success. It is true that a successful marketing programme will attract clients to our business. However, an effective public relations strategy will keep them with us, and it can even attract more customers in the long run.

12.10 In Short, What Type of Strategies Did You Focus to Become Customer-Oriented Business?

Firstly, we must perform research activities to identify latest consumers' trends and preferences. Then, we should conduct competitive market analysis. Prior to defining the marketing mix, complete research on distribution is required. Subsequently, we need to examine the financial analysis before carrying out the final strategy review and revise the research as needed. Specifically for the company finance, we will firstly identify our business plan opportunities and thereafter perform financial planning. Following that, we will strategise on investment funds and monitor our investment portfolio. The ratio of loan to total assets should be at reasonable metric. A year-over-year decrease in the metric depicts that my company is progressively becoming less dependent on debts to grow our businesses. Meanwhile, I also review my company performance regularly by performing SWOT (Strengths, Weaknesses, Opportunities and Threats) analysis. The results from this evaluation will absolutely assist us in managing strategic planning and improve on business efficiency.

12.11 Why Do You Think the Buyers Will Purchase Your House?

We build high-quality and reasonably-priced property units with remarkable design as well as perform the handover on time. We let our products to speak for themselves. The design and quality of our housing products are unique, practical and compelling that the acceptance comes from looking at the final product alone. Our property units provide a new perspective of modern design, characterised by quality and functionality with a reasonable price tag. Indeed, it is in tandem with our vision as we strive to deliver top-in-line yet affordable property units with quality workmanship which augur well for the broader economy and attract strong customer demand.

In terms of pricing, we notice that consumers nowadays are very sensitive with price. If the price is acceptable and reasonable, they will certainly proceed with purchasing the house. The buyers will look into our sales materials and make the decision. For your information, the average price for terrace houses in Batu Gajah was around RM250,000 2 years ago. But, right now, the market price has already gone up to about RM280,000–300,000 per unit for similar size of houses. It is due to escalating cost of land, building materials and labour.

12.12 How Do You Create Firm's Reputation?

Organisational reputation is associated with branding and image. Branding is essential to every business, and we implement branding by practising consistency in constructing superb quality yet affordable property units. In fact, all our projects are completed on time within its budgeted costs while maintaining quality in its final products. This kind of approach will fuel customer loyalty and create robust profit margins for our businesses in the long run.

As for Sri Anika, we aspire to build not only quality products but a quality product that will take into consideration all risk factors and deal with the root cause. We are confident that our products meet the buyers' need for their short-, middle- and long-term investment interests. Hence, we will be doing well once we provide creative and innovative products as well as houses that meet buyers' needs. We believe that this kind of approach will create outstanding reputation among local households.

12.13 What Type of Innovations that You Have Applied on Your Housing Projects?

We practise innovation in constructing affordable property units by emphasising on sustainability, flexibility and diversity of design for all market segments. For instance, we use sustainable materials in our projects such as ecobricks that have advantages in environmental friendliness over conventional bricks. We strongly believe that properties built with ecobricks are strong and durable.

12.14 Could You Please Tell Us on the Funding Capital During the Early Stage of Inception?

Our working capital is derived from both internal and external sources, i.e. self-financing and bank loan. With regard to our land acquisition, while some lands were inherited from my late father, we still continue to assemble and hunt new land bank for our future outlooks.

12.15 Have You Diversified Your Business into Other Industries? How About the Overseas Market, Any Plans to Tap into Property Development Abroad?

Our group of companies have diversified to other businesses including construction, logging, sawmilling and biomass in order to strengthen our companies' financial health. By and large, we are optimistic about our growth in the local market, and we want to focus on Malaysia for now. Creating a brand name and recognition for our work are of utmost importance.

12.16 How Do You See the Importance of Economic Factors in Contributing Success Towards Your Venture? What Do You Think About the Roles Played by the Government in Assisting Your Business?

The economic factor stands a very important position in contributing overall success of housing development projects. Pragmatically, developers tend to build different types of properties in different economic scenarios to secure their companies' interest. In addition to that, all housing development projects involve a lot of interactions with government departments and agencies. These different government departments and agencies serve their own specialised functions on application processing and approval of activities. Despite the lead-time schedule being set for each of different procedures, some applications may end up being processed in longer time frame due to unforeseen reasons. Hence, the support of government is vital to speed up the end-to-end approval process.

The housing business in Malaysia can take about 4 or 5 years to completely finish the entire processes. It starts with land acquisition, through purchasing our own land or joint venturing with the landowners. Otherwise, we may apply land from the state government. This phase consumes about half to 1 year to finalise the sale agreement. Then, before we proceed with submitting the plan to local authority, we need to work closely with the consultants especially the town planners. They should advise the housing developers regarding the layout of the houses. In fact, all the engineers of the said departments will give their opinions on our design. For example, they will advise us on the environmental factors such as traffic and pollution matters. Undeniably, there are so many government departments involved at this stage. Consequently, certain technical applications may take up to 2 years to be approved. That is why only at the third year that we will be able to start the construction and marketing. Hence, for one project, 4–5 years of effort is quite common. In addition to that, we are obliged to 2-year defect period with the buyers. I would say that to complete the whole cycle acquires 7 years of gruelling until the end of defects period.

Based on my personal observation, due to current economic uncertainty, I think we should concentrate on affordable houses. In Perak, we are looking at the terrace housing segment because we believe that the majority of local buyers are within that fragment. No point for us to aim high profit margin with the very expensive products, whereas at last, no one will buy our houses. We do not mind to lower down our profit margin as we rely upon win-win situation between the housing producers and the homebuyers as long as customer satisfaction and our company's interests are secured.

12.17 Any Awards/Acknowledgements from Industry that You Have Obtained Throughout Your Career as a Housing Developer?

I was conferred Darjah Sultan Ahmad Shah (D.S.A.P.) which carries the title of Dato' by His Royal Highness, the Sultan of Pahang, on 24 October 2015. Besides that, one of the major Chinese newspapers called *Nanyang Siang Pau* had featured my success story in its national full-page write-up on 9 December 2015. Moreover, I have been interviewed by different major Chinese newspapers for opinions with regard to housing development issues from time to time. In terms of organisation, Sri Anika Group was also the recipient of the Golden Bull Award from year 2006 to 2009. This award is acknowledged as the most prestigious recognition, organised by *Nanyang Siang Pau*, to provide an effective platform to benchmark successful SMEs and to inspire more SMEs to strive for excellence and accomplishment.

12.18 Could You Provide Any Extra Lessons that You Have Undergone as an Eye-Opener for Other Fellow Industry Players?

I was once approached by a broker who offered me to buy a piece of land at a strategic location. Due to delay in decision-making, the land was eventually sold to another third party. That land was thereafter developed by another developer and sold to homeowners with brilliant return. Therefore, I would like to share my lesson learnt as, "never hesitate in making prompt decision to acquire potential land even though it is pricier than market price". That spirit must be kept alive and the torch of the struggle should be kept lit at all times.

On a separate note, I would love to touch a bit on the current challenges of the industry. There will be unexpected tough issues that crop up in any housing project. For instance, there is a possibility of technical fault, squatter issue, problem with access roads (e.g. trespassing private land, acquisition by government), imposition of building design by local authorities (especially at royal town or historical city/town), restriction to demolish old historical building, complaints by neighbours, etc. Obviously, housing developers need to stay focus and act fast to resolve these issues effectively, or it can potentially affect the project's deliverables.

Part III
Large-Scale and Public-Listed Housing Developers

Chapter 13
I&P Group Sdn Berhad

Keywords I&P Group Sdn Berhad • Jamaludin Osman • Chief Executive Officer • Township development • Leadership • Strong branding

I&P Group Sdn Berhad is indisputably one of Malaysia's leading property development companies. Its trademarks include quality products and high design standards, qualities that have long been recognised by both its peers and its clientele. The company has seen many changes in its journey to the list of top ten developers in the country. Established in 1963 as Island & Peninsular Development Ltd., it underwent a name change to Island & Peninsular Development Bhd in 1984, the year it launched its first township, Taman Setiawangsa in Selangor, as its only developer.

In 2004, as part of the rationalisation exercise by its majority shareholder, Permodalan Nasional Berhad (PNB), Island & Peninsular Berhad acquired Golden Hope Plantations Sdn Bhd, I&P Seriemas Sdn Bhd, I&P Alam Impian Sdn Bhd and Syarikat Perumahan Pegawai Kerajaan Sdn Bhd (SPPK) as part of the rationalisation isometrics making it the largest property developer in Malaysia in terms of its land bank. In April 2007, PNB proposed a voluntary general offer to acquire the remaining RM285.3 million shares, and the company was subsequently delisted voluntarily from the local bourse in July 2007. In May 2009, a new name in property development, I&P Group Sdn Berhad, arose from the successful merger and rationalisation exercise between Island & Peninsular Sdn Bhd, Petaling Garden Sdn Bhd and Pelangi Sdn Bhd. This union brought with it a long and prestigious track record in the property business, with the merged entity's assets amounting to more than RM10 billion.

I&P Group Sdn Berhad is currently helmed by Dato' Ir Jamaludin Osman, whose history with the company goes back a long way. Previously appointed as the group Managing Director of Island & Peninsular Berhad in 2004, Dato' Ir Jamaludin Osman has continued to implement his astute strategies upon assuming the position of group Managing Director of the merged entity in 2009. Having risen up the ranks from the position of civil engineer to Chief Executive, he is well placed to mentor and lead his people towards excellence, as he has a broad understanding of the

M. Jaafar et al., *Business Sustainability Model for Malaysian Housing Developers*, DOI 10.1007/978-981-10-5266-8_13

different functions required in property development. His extensive experience in civil engineering, management and marketing spanning three decades in the property industry, along with his motto *We Can Do It,* has helped him take the company to new heights of achievement and won it numerous accolades in the form of industry awards.

He continues to put his heart and soul into meeting the needs of I&P Group Sdn Berhad's clientele, taking a personal interest in ensuring that the quality of the properties delivered to house buyers meets his exacting criteria. He thinks nothing of dirtying his hands at a construction site if he has to, in order to ensure his standards are fulfilled. He has a unique strategy for identifying buyers' desires – he puts himself in their shoes to identify their needs and wants in terms of quality, design and service and strives continuously to inculcate this in his team, telling them to 'build as if the house belongs to you'. His customer-centric philosophy has helped the group realise its aspiration of building sustainable communities that exceed expectations in terms of quality and timeliness. Today, the group has a stellar reputation for developing premium-quality residential projects with unique concepts or themes. Dato' Ir Jamaludin shares with us his invaluable experiences garnered over 30 years. Here are excerpts of the interview:

13.1 Would You Mind Sharing Your Story with Us?

I am really just a kampong boy; I am not from an elite background. There were just four of us in my family – my parents, my younger brother and me. My father was a rubber tapper in Temerloh, Pahang, while my mother was a housewife who sometimes helped my father tap rubber in order to make ends meet. My father was a disciplined man and he was very strict with us. I am a better person today because he brought us up to be hardworking and to think positively. I guess the 'push' factor that made me who I am in the corporate sector today was my desire to serve my country. It was my childhood dream to be a civil engineer, and I realised that was the way I could best contribute.

I am a registered professional engineer with the Board of Engineers Malaysia. I graduated with a Bachelor in Civil Engineering from Universiti Teknologi Malaysia in 1979 and began my career with the Pahang Tenggara Development Authority (DARA), opening small townships such as Bukit Ibam, Keratong and others in the state. I subsequently worked with Dewan Bandaraya Kuala Lumpur (DBKL) and then with PETRONAS as a senior civil engineer. However, the role of a civil engineer at the petrochemical plant was not challenging, and I decided to move on to the property development industry as I thought it would better fit in with my passion for civil engineering.

That was how I began my involvement with property development. I joined SPPK in September 1985 and was involved in construction, supervision and the other managerial tasks that a civil engineer is involved in. The big transformation came for me when I was assigned to head the marketing division in 1992.

It was a great paradigm shift for me, transitioning from the technical side of property development into marketing and sales. To meet the challenges of my new role, I attended a management development programme at the Asian Institute of Management in Manila in 1992, and this gave me a host of new ideas. One of them was to enable the marketing team to spearhead product design, so that we could produce houses that were relevant to market needs. This became a philosophy we subsequently adopted to this day in I&P Group.

13.2 In a Succinct Manner, Could You Please Explain What Your Position Involves?

As the Chief Executive of the company, in my opinion, the CEO's role is crucial in ensuring the company's performance. My leadership style determines the support I get from my team, and a company's performance depends very much on its people.

As the CEO, I carry out my duties with full integrity to ensure the best for all stakeholders. I also need to set a clear vision for the organisation and to lead my people into realising that vision to meet stakeholders' expectations.

13.3 What Are the Strengths of I&P Group if We Compare with Other Real Estate Developers that Are Predominantly Chinese-owned?

I&P is a township developer that believes strongly in giving quality of life to those who buy our properties. We have a track record for developing big townships, which include Bukit Damansara, Bandar Kinrara, Alam Impian, Alam Damai, Alam Sari and others, in which we provide high-quality homes at affordable prices, together with the amenities necessary for a comfortable lifestyle.

We have continuously stressed on quality and timely delivery of the homes to the purchasers, and I am happy to say we have consistently delivered properties to our buyers on time or even earlier than stipulated. That is one of the many reasons why we have successfully gained and retained the trust of our purchasers as well as establish our reputation as a reliable developer that delivers on its promises.

13.4 You Are Managing a Number of Companies Under I&P Group. What Are the Unique Strategies that You Employ to Manage Them Well?

One of the most important strategies is to be well-prepared ahead of time. For example, the challenge for many developers is to get the approvals on time, so we can launch our projects as planned to enable us to meet our business goals. At I&P, we work to get the approvals as early as possible, which allows us to have some flexibility with deadlines that would enable us to react quickly to policy changes by the authorities or changes in the market demands.

Another vital factor for success is teamwork. Leaders can never achieve the vision alone as they need to get the buy-in from their people to ensure that everybody is aligned with the company goals and is accountable for their own scope of work. These are achieved through regular projects and coordination meetings to keep everyone in the loop. We have a diverse technical and administrative team that will ensure that the different policies of the various states where our projects are based, are being tracked and followed closely. Through concerted communication, any new policy requirements are then updated so that everyone will be kept abreast with the various policy changes.

13.5 What Is the Role of the Board of Directors in This Company?

The Board of Directors normally discusses the strategies of the company during board meetings. Besides that, the Board will be informed on the progress of every project and the challenges faced by the management. The Board also reviews and approves any proposals submitted by the management.

13.6 As a Civil Engineer Who Has Become a Managing Director, What Strategies Have You Implemented to Succeed in Your Career?

You need to be focused and be willing to learn. Learning does not stop even if you are a Managing Director, and you must be willing to learn from people who have more experience in the industry. You need to read and apply smart learning strategies and continue to enhance your competencies. I had attended a 2-month management development programme in Manila which has helped me to understand the many aspects of running an organisation.

My experience as a civil engineer in property development provided the exposure on all aspects of construction. Until now, I still continue to carry out inspection at the project site. In short, in order to be successful in your career, you must be willing to work hard, continue learning and always be sincere in carrying out your duties.

13.7 What Are the Extraordinary Characteristics of I&P Projects? How About the Overall Achievements in Terms of Projects?

I&P Group Sdn Berhad is best known for its township development. Building homes with a heart and putting customers first lie at the nucleus of our culture, so customers trust I&P to always deliver on excellence and innovation. The vision of propelling I&P to greater heights has always been backed by a mission that clearly puts the customers' interest first. This emphasis has helped I&P to emerge as one of the most respected housing developers in Malaysia.

The company has a land bank of 1437.44 ha concentrated in the Klang Valley and Johor Bahru with a potential gross development value of more than RM25 billion. The land bank is expected to keep the company busy over a span of 15–20 years. It has a portfolio of 14 township developments spread across the Klang Valley and Johor Bahru. The 14 townships are Alam Damai in Cheras, Bukit Damansara and Bandar Baru Seri Petaling in Kuala Lumpur, Bandar Kinrara in Puchong, Alam Sutera in Bukit Jalil, Alam Impian and Temasya Glenmarie in Shah Alam, Alam Sari in Kajang and Bayuemas in Klang. They also include Taman Pelangi, Taman Pelangi Indah, Taman Perling, Taman Rinting and Taman Industri Jaya in Johor.

New townships to be developed by the group which are currently in various planning stages are being planned in Selangor, where we own 161.47 ha in Meru, Klang; 38.04 ha in Brogaville, Semenyih; and 58.68 ha in Salak Tinggi, Dengkil. In the southern region, we own 388.09 ha in Ulu Tiram, Johor Bahru, and 137.19 ha in Gelang Patah, Johor Bahru. The profit before tax of I&P Group is RM278.58 million in 2011, RM378.24 million in 2012, RM526.14 million in 2013 and RM442.73 million in 2014.

13.8 From Your Viewpoint, What Are the Measurements Used to Gauge Success in the Housing Industry?

The first is profitability and then its strong branding, or image, that plays an important part in achieving success. Other than that, the sustainability of both company and development projects are among the vital criteria. You have to develop the opportunities that will establish you as a credible developer. That is why we are very focused in terms of supervision, project location, pricing and product delivery.

Success is also dependent on quality since this is associated with the name of a development company and its projects. We aim to influence public perception towards I&P by providing on-time delivery and acceptable quality homes with a high capital appreciation in good locations. The positive feedback from buyers is crucial for us to boost our sales and performance.

13.9 Which Department Would You Say Is Crucial to Ensure the Company Excels in Its Performance?

I think all departments are important because we work as a team. To give you an illustration, the marketing department is essential because building houses without selling them is rather impractical. Similarly, the finance department ensures that we have funds for our projects, and the credit team ensures smooth sales collection is important, too. On top of that, the human resource team plays a vital role in keeping the people on track in the organisation.

It is very difficult to say which is more important as they each have their roles that are needed to run this company. I would rather classify them into two categories: support and operations team. The support team are those who monitor the company's performance from the office, while the operations team are those who execute the projects.

13.10 How Do You Tackle and Control Business Risks?

There are many risks in housing development and that has pushed us to establish a risk management team. Regular meetings are held to discuss the possible risks that we are exposed to, and the solutions to mitigate these risks are being discussed and endorsed during these meetings.

13.11 Sustainability of Housing Development Firms Depends Very Much on Their Land Bank. What Is Your Opinion on This?

We are always on the lookout for additional land banks that are viable for our future investment. We look for the best location and the acquisition cost of the land bank and assess its marketing viability based on our set of criteria before we commit. Another important aspect is the land bank's readiness to build. If the timing is right, aspects such as technical conditions of the land should be considered (hilly then cut and fill; soft and soggy then vertical drain and overburden pressure). All these elements have impact on the layout and building approval. Taking too long will cause it to miss out on economic cycle.

Buying land is not like buying a house. We do not have a specific target, but it is most likely to be in or near the Klang Valley. To get a piece of land in a good location like the Klang Valley is hard. That is why we predict that within 10–15 years, the development trend will go upwards into strata properties – no more sprawling terrace houses – or perhaps, landed properties will be very much a premium item in the future. Within our development plan, we are also initiating high-rise projects that would enable us to expand the life span and profitability of our land banks.

13.12 The Growth of Science and Technology Also Has Great Impact on the Housing Industry. How Does I&P React to This, Especially in Terms of Building Innovation?

Everybody is looking towards being IT savvy. You may notice that people are talking about security applications, CCTV implementation and other facilities of that ilk. In terms of building innovation, we introduce simple green elements in some of the houses we build, as we think to make it a fully green building, it will be costly as well as taxing on the homeowner. Not everyone is ready to maintain a green building.

We try to keep it balanced. In our semi-detached and bungalow projects, for example, we put solar heaters and a rain harvesting system. We have no plans to make our houses fully green as it will indirectly increase the construction cost. We innovate by offering better finishes and more durable products. Advancement in technology has helped produce building finishes such as tiles, toilet fittings and door accessories to be of better quality and design. We are given more choices which offer many new composite offering natural materials resulting in natural and more durable products suitable for our climate, thus enhancing the overall look of our end products.

13.13 Do You Find Networking to Be an Indispensable Feature in the Firm's Success?

Networking is imperative, not only with the local authorities but also with our peers and fellow industry players. We can share problems and success stories with them as it helps us perform better. As a member of the Real Estate and Housing Developers' Association (REHDA), we get first-hand information especially in relation to government-related issues.

13.14 How Does the Economic Climate Affect the Performance of Your Company?

It is applicable to all businesses, not only the housing industry. When it affects the national economy, the consequences will be greater for the housing industry. The property cycle is always there and we have experienced the recession period. As long as the economy is fine, with steady growth, the housing development industry will certainly fare well. But, if it goes downwards, many other businesses will also be affected. Sales are sometimes hindered by the strict financing criteria and the 30% real property gain tax (RPGT) that makes investors think twice before investing for short-term gain.

Personally, these are just obstacles for us to be more innovative in our marketing strategy. We continue to launch projects and manage our cash flow by reducing the numbers of unit launched. This makes sense because in the event of a slow market condition, the situation will exert less pressure on our cash flow. For me, this daunting task called for hard work and financial prudence.

13.15 In Terms of Market Focus, What Types of Houses Do You Concentrate on Building?

We develop many types of houses: low-cost, terrace houses, semi-detached and bungalows. We focus on various types of homes depending on the demand in each development.

13.16 Who Are the Buyers of Your Housing Products?

Usually, it is a combination of Bumiputera and non-Bumiputera purchasers, comprising first-time homebuyers, upgraders, those who wish to own bigger houses and investors. The majority of our developments are freehold in status.

13.17 Do You Implement Outsourcing or Internal Recruitment in Terms of Staffing?

We engage consultants to prepare the tender documents; our internal staff comprising of quantity surveyors, architects and other related professionals are entrusted to check and assess the work at the preliminary stage.

Designs are also outsourced to external architects, but the concept, design guide and theme direction are under the close scrutiny of the internal product design team. This enables us to focus on what matters and help speed up the work processes.

13.18 What Are the Major Selling Points of Your Projects?

Location, quality, on-time delivery and value creation. Most of our projects are situated close to the city centre and have excellent highway connectivity with ample amenities provided by the township itself or within its neighbouring vicinity. We are also well known for our quality finishes as well as on-time delivery or ahead time of delivery. Capital appreciation is another factor enjoyed by the buyers of our products. Buyers have seen a capital appreciation of more than 50% of their initial investment.

13.19 Do You Mind to Share with Us Any Recognitions or Awards Received by Your Ownself and I&P?

Due to my success and contribution to the corporate arena, I was conferred a Datukship in 2005 and was named 'Property CEO 2014' by FIABCI Malaysia. I was also recently named the 'Personality of the Year' by *Property Insight* (Malaysia) magazine in 2015.

On a separate note, for 11 consecutive years since 2003, the group has won The Edge Top Property Developers' Award. It also won The Edge-PEPS Value Creation Excellence Award 2010 and 2011 and was the merit winner in this category for its Temasya Glenmarie project in 2013.

Anyway, our latest awards in the past 2 years include:

- PNB Bumiputera Development Award (2014)
- PNB Dividend Contribution Award for the highest dividend in the last 5 years (2014)
- The Brand Laureate Best Brands' Signature Award in Real Estate Property for 2014–2015
- Best Public Services Development Malaysia for Bukit Damansara Public Library (Asia-Pacific Property Awards' World's Best 2014–2015)
- Highly Commended Award for Commercial Landscape Architecture Malaysia category for Alam Impian Park in Alam Impian (Asia-Pacific Property Awards' World's Best 2014–2015)
- Global Leadership Award 2015 for Excellence in Property Development Sector by The Leaders International

- Global Branding Awards 2015 for Global Best Brands Township Development MNC Category by The Leaders International
- A Highly Commended Award for Architecture Single Residence Malaysia from Asia-Pacific Property Awards 2015–2016 for the Temasya Glenmarie project

13.20 Do You Have Any Last Words of Wisdom to Share with Other Industry Players?

To be successful, you have to work hard, be honest and sincere, be contented and practise good values in your work. My personal life motto is simple, *I can do it,* because the moment you adopt negative beliefs, things will not work anymore.

Chapter 14
LBS Bina Group Berhad

Keywords LBS Bina Group Berhad • Lim Hock San • Family business • Courageous • Community involvement • Media relationship

Before being listed in Bursa Malaysia, LBS Bina Group Berhad used to be a small family-owned business, founded by Dato' Seri Lim Bock Seng in the 1960s. It started as a construction business in Petaling Jaya, Selangor, before venturing into property development in 1992 with its first project, Jelapang Maju Light Industrial Park at Jelapang, Ipoh. Subsequently, LBS delved into housing and township development after the Founder's son, Tan Sri Lim Hock San, took over the business. Under his helm, LBS have moved into commercial and industrial development as well. This trend of growth has seen LBS become a dominant player in Malaysia's property sector and spread its wings across the seas to China. In the 2000s, LBS expanded into related activities such as investment holdings, property management, turfing and landscaping, insurance agency and building materials.

Several of its well-publicised accomplishments were Taman Perindustrian Bukit Serdang Seksyen 14 in Selangor – jointly developed with players from the 1992 Thomas Cup champions, Taman Pinggiran Putra, Serdang, and the township project at Bandar Saujana Putra. Among the recent flagship projects are D'Island Residence, Puchong, Bandar Putera Indah, Batu Pahat and Sinaran Mahkota in Kuantan, Pahang. LBS is constantly aware that affordability and rising house prices remain the major concerns among Malaysians. As a community and socially responsible developer, LBS will continue its effort to deliver quality and affordable homes while taking into consideration the requirements pertaining to stringent home financing, high-interest rates, increase in building materials and land cost, property cooling measures and current 6% of GST.

M. Jaafar et al., *Business Sustainability Model for Malaysian Housing Developers*, DOI 10.1007/978-981-10-5266-8_14

14.1 Could You Please Tell Us on How This Company Was Founded?

Like all success stories, LBS began with one man's vision. In the 1960s, my father, Dato' Seri Lim Bock Seng, founded a construction business in Petaling Jaya, Selangor. Prior to that, he opened a coffee shop. He has more than 30 years of experience in the construction and property development sectors. He was initially involved in the supply and transportation of sand, aggregates, earth and ball clay for construction of infrastructure works such as earthworks, sewerage works, roads and drainage. Subsequently, he ventured into construction of houses, shops and factory buildings. Now that he has retired, his sons have taken over the mantle in the firm. After a few decades of rapid growth, LBS sets its sights higher and ventured into the real estate sector in 1992 where we started our first project of Jelapang Maju Light Industrial Park at Jelapang, Ipoh. Within a decade, through numerous successful projects, LBS becomes established as a reputable developer.

14.2 May You Elaborate a Bit on Your Family Background?

I have ten other siblings with four elder sisters above me and I am the first son. I have three younger brothers. My father only had formal education until Standard 3. His four sons, including me, and the youngest sisters were able to get opportunities to pursue our studies overseas. He worked hard to ensure that our educational expenses were adequate in order to get recognised degrees.

Frankly, we had quite a challenging time studying overseas. I graduated in 1982 and returned back in 1983 to work with my father in his contracting firm. During that time, I saw that contractors were always bullied by the developers with a lot of payments uncollected. I believe you are aware that contractors work for people and sometimes hardly get payment for their works. Thus, I said to my father that I wanted to be a developer.

In 1990, after the recession, I suggested my father to venture into development. When we become a developer, we made sure we treated our contractors well because we knew of their hardships. If we do not pay them, the works will obviously be delayed.

Hence, in 1992, we changed our business to property development through our first project with players from the champions of the Thomas Cup team. The Selangor State Government had given ten pieces of land to them (each of 5 acres) that involve the Sidek brothers, Foo Kok Keong and other team members. Luckily, they chose us as the developer for an industrial development in Bukit Serdang.

This is a family business and involves mostly my family members. Currently, my father is the Chairman of the company since 6 December 2001.

14.3 How About Your Personal Background?

I have a degree (first-class honours) in civil engineering from the University of Wales, United Kingdom, graduated in 1982. I got married during my sophomore and have four children, three of whom have married. Two of them are working with me, while the others are housewives. During my tertiary education in the UK, I won the second prize in the British Steel Corporation Competition for Design in Hollow Steel Section in 1982. Personally, education is merely a platform for me to strive for a better life since it only provides the basic things. Most importantly is how we are going to serve the society. Besides doing business, I am very active in community development as I sit on the board of various charitable organisations.

I believe that through my excellent business foresight, technical expertise and 33 years of management experience have positioned me well as the key leader. For your record, I was also the recipient of the inaugural the Platinum Entrepreneur Award by SMI Association of Malaysia in 2011, the prestigious Entrepreneur of the Year 2012 by Asia Pacific Entrepreneurship Awards (APEA) and Malaysia Best Company for Leadership of the Year 2014 by International Alternative Investment Review (IAIR) Awards for my outstanding entrepreneurship, leadership and achievements in raising the profile of the company in the country and abroad.

14.4 What Is Your Vision in Life? Do You Set Any Personal Target in This Business?

My vision in life is very simple. The first thing is your family must be happy. I put my family as my first priority in every aspect of life. Personally, I want to build my own legacy that creates something different for LBS in the future. Business-wise, we want to make sure that LBS will be able to become a quality homes provider. Until now, I think we have performed quite well since we received a number of recognitions from both the public and government.

14.5 Do You Have Any Business Mentors that Have Inspired and Assisted You in Business? What Were the Lessons that You Have Learnt?

I would regard my father as my business mentor because he had taught me a lot. He said that whatever we promised, we must try to fulfil. We shall not cheat people and must be sincere and honest in every deal even without black and white proof. Do not take people for granted. When I joined the company, he was already the Chairman of the firm. He always asked me not to quarrel with my brothers. That is the usual reminder from him.

Besides him, I will consult with my brothers and fellow friends, asking them on the opinion to gather exact information and for further verification. It is true that different people render different views. After all, in handling the business, it all depends on you, not on anyone else.

14.6 How Do You Define Success in This Housing Business?

As a private property developer, we inherently want to make money. People nowadays prefer to buy a home that signifies a luxurious shelter for future investment, with certain concepts to be fulfilled such as garden concept, guarded concept, etc. With such views, people will definitely buy the houses as they predict the values will go up. The existence of homebuilders like us has created value for homebuyers through our renowned brand.

Take the example of our flagship project, Bandar Saujana Putra. We started with low-cost and low-medium-cost products as we poured and spent money to build the interchange and other infrastructures. Today the township is very vibrant. Besides that, we usually interact with residents through my public relations staff. If they convene any community activity, we will participate and provide sponsorship, while if there are any problems raised, they will come to see us. We are a responsible developer; we encourage the homebuyers to jointly promote their residential area, so that the value of their houses will be improved. It is better for them to work together with us to uphold the brand name of the residential area.

Bandar Saujana Putra is a new township with more than 90% of development have been completed, and we still have a balance of 70 acres for commercial development that will comprise apartments, shoplots and others. About 60–70% of the residents are Malays and Bumiputeras. We will assure to complete the projects. That is our promise. Within 2 years, we will try to finish the delivery.

14.7 What Do You Think About ‘Luck’ in Handling Businesses Especially in the Housing Sector?

Luck is for those who are well-prepared. Luck may bring fortuitous gain to you, but we must never depend too much on luck. Even if we have better luck, we need to fight for the best through our tireless effort. Opportunity knocks our door only once; we must grab it, or otherwise we will lose it. Do not miss the chance to keep growing. This luck does not come often, as it only appears once or twice in your life.

14.8 What Are the Traits that Should Be Instilled by a Businessman Like You in Order to Succeed in This Industry?

There are few attributes that come to my mind. You should be humble and people-oriented, especially towards business friends. If we wish to have their respect, we must first show our good manners to them. Also, try to be honest in terms of helping people, so that people will respect you. Apart from good leadership and profound passion towards the business, which I perceive as common in business undertakings, we have to open our mind-set with the ability to adapt to market changes. Also, businessmen need to be wary at all times. I recommend my fellow entrepreneurs to take one step at a time. Not only hard work is pertinent, but commitment towards work is very important. I trust an old adage that 'as you sow, so you shall reap'. Most importantly, no decisions should be made when you are emotional, be it angry or happy, and you must give time of 2 or 3 days to ponder upon for the best decision.

14.9 Please Tell Us the Managerial Side of This Company. Perhaps You May Share with Us Any Specific Methods Utilised in Daily Form

In our organisation, we develop an effective system that I noticed to be a great approach for the management of this company. Through this ideology, we reckon that the whole is greater than the sum of its parts. I would chair a meeting every Monday with HODs. Normally, I will spend an hour with them. I share knowledge with my staff, and if you perform well, we will reward you. Since we are into the niche market, I have shared with them the company's policy and strategy. Everybody has their own department and responsibility. All departments are important as they are interconnected. It is like a system: one section breaks down, others will be affected. Everybody has their own share in this company. Henceforth, we educate them to make sure that they are committed with their works.

It undeniably takes time to see the transformation, but for the past 5 years, we have improved a lot and our staff have understood our working culture. At the end of the day, everybody gets bonuses. I practise a moderate approach, not to be so tense towards my staff and flexible with environment and conditions. We acknowledge the requirements of a competent team to beef up our organisation, be it from construction site or management office.

14.10 Do You Think This Property Business Is Risky?

Actually, any business is risky. There is no such thing of business without risk. But we believe in the philosophy that 'high risks, high gains' and 'no risks, no gains'. The company should be able to control the risks. Last time, when we were small, we can go into everything, but now, as we are getting bigger, we need to go through our risk department, and if they doubt or persistently disagree, I may withdraw my suggestion and think about other channels to continue. We also need to heed our cash flow since we are dealing with too many high-rise projects; we are changing some of the concepts to landed properties, as our profit recorded lesser margin. It is very important to control the cash flow. If cash flow goes wrong, everything will be affected. Again, we are proud to tell you on the effective system that we are applying with since LBS is not new in the real estate market. With more than 30 years' experience, we have gone through major economic crisis, such as in the mid-1980s, 1997 and 2008.

14.11 Why Do You Want to Make Your Firm as a Public-Listed Company?

Frankly, we encountered a lot of constraints to get listed. But, if we are listed, we can raise money from the market. When we approach landowners for JV, people trust you more. They will have confidence that we will be able to complete the projects till the handover period. Listed companies are more exposed to the media, and sometimes, we are invited by government agencies to participate in JV privatisation, enabling us to gain more contracts and contacts. For your information, we went for listing in Bursa Malaysia in 2002. We have few hundred acres of land, and by listing our company, the cash can be easily accessed to further expand through wider resources. But for us, we are a bit conservative and wary in our movement to acquire more businesses. We make sure that our shareholdings will be able to maintain certain percentage of ownership.

14.12 Regarding Your Company Activities, Who Are the Decision Makers Accountable for Organisational Assets and Projects? What Type of Housing Projects Your Company Is Concentrating on?

Aside from me, we allow my other brothers to make decisions. As far as we are concerned, any decisions that benefit the company will be accepted. At the moment, LBS will be focusing in affordable to medium residential projects in 2014–2015. I think we have completed many projects, and I would like to stress that we have never abandoned any projects even during difficult times. We also have projects in Pahang as we are focusing our development in Cameron Highlands along with new ventures in Kuantan for commercial and housing properties. The projects are at the infancy stage and still ongoing.

14.13 Could You Please Tell Us a Bit on Your Experience Developing Houses in Cameron Highlands?

We are currently the largest developer in Cameron Highlands and committed to the longer-term development there. LBS has made a strategic move into Cameron Highlands as early as 1998. I am pleased to say that as of today, we have a total of 223 acres of land in Cameron Highlands for ongoing and future developments worth about RM5 billion in gross development value (GDV). This also indicates that our developments in Cameron Highlands contribute about 26% to the overall group GDV. We had successfully completed two projects, i.e. Carnation Park and Taman Royal Lily, through a joint venture strategy with the Pahang State Government. The third one is Cameron Golden Hills, a joint venture project with Pasdec Corporation Berhad, while the latest project deals with the federal government. We are set to inject leisure and entertainment elements to our future developments to spice up the overall appeal. The upcoming Cameron Centrum will be a project involving six precincts with a GDV of RM2.7 billion.

14.14 On a Separate Note, How Did LBS Manage to Acquire Business in China?

These properties belonged to Instant Green Corp Bhd, a listed company in Malaysia. Due to the financial crisis in 1998, the company went insolvent. While we were also interested to go for the company's listing, we did a reverse takeover. We took over this company through a bidding process from Danaharta. Coincidentally, they have projects in China (i.e. Zhuhai International Circuit) but with myriad of financial and managerial problems. After so many years of delays, the problems were eventually resolved. Our current Prime Minister, Dato' Seri Najib Tun Abdul Razak, who was the Deputy Prime Minister back then, helped us a lot to expedite the process with the government of China.

14.15 What Factors Do You Think Have Made LBS as One of the Leading Developers in Malaysia?

We have more than 300 staff which I consider as our greatest asset. Every home that we develop and deliver bears the essence of our mission, which is to develop inspiring spaces to enrich communities. I would dedicate this success to the committed LBS team and our stakeholders including the homebuyers who continue to support us.

For housing developers, homebuyers are important as they absolutely want better quality workmanship. That is why we have an in-house construction firm to assist our own development. If any purchaser makes a complaint, we will entertain every single report. Buyers nowadays are very smart and demanding, to a certain extent that they are private messaging me via Facebook, specifically regarding our service and products. As they tend to lodge reports and complaints to me, I will subsequently channel them to my staff in charge. We will make sure that our promises and their needs will be fulfilled. We would not compromise on poor quality in order to gain confidence from buyers.

It is appealing to note that we are still connected with our homebuyers through the residents' association and support the activities organised by them. I perceive that we need to treat them as friends so that they will feel as part of our big family. It is a win-win circumstance as it will also enhance the value of township and our reputation.

14.16 That Being the Case, Why Do You Think the Buyers Are Purchasing Your Houses?

The name and brand play vital roles. Aside from that, we also pay particular attention on the current competitive price, location, accessibility and housing concept. Our commitment towards serving the buyers is always a top priority. Project-wise, for BSP Skypark in Bandar Saujana Putra, we are confident of attracting both first-time homebuyers and those who are seeking to upgrade their status to more luxurious home selections. In fact, there is good demand for our medium-range developments. The easy access, strategic location and fast maturing townships appeal to first-time homebuyers. Meanwhile, LBS also caters to the luxury segment with our inspiring concepts and living spaces at D'Island Residence. These offerings appeal to the affluent market since it features diverse modern qualities with an emphasis on young and first-time housebuyers. We continue to push the boundaries of the property industry to truly create living habitats of the future.

14.17 So, How Does Your Company Retain Your Buyers?

LBS is a customer-oriented company. We offer products which create value of money for homebuyers. Housing developers need to be friendly; perhaps they should set up customer service and maintenance departments to safeguard the welfare of buyers. Whatever problems arose, we need to tackle and try to resolve them accordingly to please our housebuyers. As I said before, we work together with our residents' association on each activity organised by them. Take the example of our flagship project, Bandar Saujana Putra; we want it to be a 'township for everyone'.

14.18 What Do You Think About the Importance of Land Bank for Housing Developers? In Relation, How Do You Implement Land Selection?

Land bank is important for the sustainability of the company. As a developer, we have to replenish our land bank. Even though we have future projected GDV, we must always seek for new lands. Both land banking process and locational aspects are of paramount importance for homebuilders like us. We need to spot and judge the right location, study the current market trend and observe the changes of land price. Normally, when we see the land, we will negotiate the price and assess the marketability of the area. What sort of target market and what kind of houses are the questions that need exact answers to be solved. From there, we can roughly know how many houses can be erected based on density, cost of the land, etc. Our marketing staff will start to implement the marketing survey on price ranges that we should consider and others of the ilk. We then begin to bargain with the landowners either to purchase directly or JV.

While we are having projects in different states, it is imperative to mention that this kind of approach is significant to spread out the risks. Besides that, the land price in Klang Valley is escalating with more competition from both local and foreign investors. Other than that, we have contacts in Johor and Pahang that facilitate us to implement projects there. An exciting fact to be shared is that other state governments are also welcoming us to venture in their territories such as Kedah, Kelantan and Melaka. I expect to build landed properties there.

14.19 Is It Critical for Housing Developers to Market Their Own Projects?

Marketing is very important and cannot stand on its own. In certain cases, we get outsiders to launch and sell our projects. As we expand and grow our business, we have also opened two new sales galleries at Cameron Highlands and Kuantan in

Pahang. We would not have ventured to this state without the support of our valued stakeholders, business partners and the management and staff of LBS.

14.20 Do You Think Network Plays a Significant Role to Achieve Success in Business?

Without doubt, excellent relationships with local authorities, government agencies, consultants and even competitors are crucial in this industry. It is the chief requirement for us to succeed. Additionally, network with media is also important. Our media exposure is quite plenty. Articles in newspapers regarding our company are great to boost our image and sales. It must be strengthened from time to time. Honestly speaking, they have been very friendly to us over these few years. Besides, we also organise 'Media Appreciation Night' once a year to cherish their role and coverage about our company.

14.21 Since Malaysia Experiences Precarious Housing Cycles, What Strides Have Been Taken by LBS to Sustain Its Success?

LBS remained resilient in 2013 and 2014 despite Malaysia's volatile property market. This is because the group was able to adapt to the needs of homebuyers quickly and effectively by providing a good mix of projects. The group's commitment towards enriching communities continues to be a key element in our developments that drives demand for LBS homes. We are confident LBS will enjoy continued growth in 2015 with new ventures and developments to be unveiled.

The Central Bank of Malaysia (BNM) has forecasted that the national economy will grow at 4.5–5.5% this year (in 2015). A strong economy indicates fast-pace movement of the industry allowing us to develop middle- to high-end properties. If the economy slows down, we will reduce the pace of development by only focusing on affordable properties.

If the gross domestic product (GDP) is consistently at 5%, the business will be able to sustain. That means our economy is faring at a more healthy stage. I believe that if lower GDP is recorded, it will create inflation. For us, the global economic slowdown does not affect much on our local housing market. With China getting dominant in world economic development, I perceive that our future outlook will be brighter. Paradoxically, we do not depend too much on the American economy.

14.22 Does Government Play Important Roles in Your Business?

The government helps us to promote the properties in Malaysia. They have promoted Iskandar Malaysia to Chinese investors. The government has done a lot, but the government should not compete with the private sector to do business. I would like to stress that our government has done a good job to facilitate the development via tourism. Besides, they have provided infrastructure projects such as highways, LRT and high-speed rails and give us some opportunities to take part in their housing activities. I think this is one of the advantages for being a public-listed company as we get more chance to tender the construction and housing jobs, while people or public can also access our market cap.

For instance, the Selangor State Government had asked us to build low-cost houses. They also gave us several parcels of land to develop medium-cost houses, which is why we have already built more than 30,000 houses. We still stick to medium-cost housing since the cost of building materials has gone up. The price depends on the location; I would say it is about RM300,000–RM500,000 per unit. The scenario is rather similar with other states. We also develop affordable houses in Batu Pahat, Johor, involving 1614 units, which are priced at RM150,000 per unit. Interestingly, we have obtained other invitations from other state governments like Kedah, Pahang and Kelantan to develop affordable homes. Indeed, while the ultimate aim is to have some profits, it may not be as much as doing the high-end properties. In this sense, the government has played their role via the establishment of PR1MA to cater for low-cost houses. As a private property developer, we will try to contribute although it may be not so much.

14.23 What Are the Plans of LBS in the Near Future?

We plan to launch our new projects in Pahang, Klang Valley, Batu Pahat, and will continue to source for strategic development lands to further enhance our performance. Apart from enlarging our land bank, as per now we have more than 1800 acres; we also wish to double the Malaysian operational profit after tax from RM37 million to over RM75 million in 2015. We are still on the lookout for potential land bank either on direct purchase or joint venture basis. Our principal business is still in property development. In spite of that, we will go for diversifications that may contribute to our profit augmentation. Perhaps, the tourism business is worth a look in. But, it is still under review, only for upcoming plan.

14.24 Has LBS Receive Any Awards for Its Outstanding Projects?

We have received numerous awards locally and internationally since the 2000s. The most recent awards were garnered during the prestigious regional property awards, Asia Pacific Property Awards (APPA) 2014. We bagged the top award for three categories, namely, the Best Residential Development, Asia Pacific, with SKYVILLA at D'Island Residence; Five-Star Best Residential Development, Malaysia, with SKYVILLA at D'Island Residence; and Highly Commended Interior Design Show Home, Malaysia, with Nautilus at D'Island Residence. A year before, we were the award recipient at The Malaysian Reserve (TMR) Property Press Award 2013 for the best urban scenic development.

14.25 Any Specific Working Cultures that Are Implemented by Your Company that You Think Contribute to the Business Success?

Through effective systems that I have highlighted before, we always employ internal briefing with the staff to share knowledge among them. I encourage my staff to tell or voice out mistakes, and in the case of major mistakes, we will punish and warn them. But we will then guide them for the corrective actions. Comparatively, if they make improvements later, they will be praised and also given some recognitions. Other than that, I will ensure that all staff are aware with the company policy and able to work together. All my staff are equally important to the success of this organisation. Another thing is every staff has the opportunity to present during our meeting to express their capability and flair. We would like to see the company's profile from the lens of their perspective.

Afterwards, I always push my human resources department to develop the internal talents. I encourage internal promotion in terms of staff development, so that we can educate and train our workforce for them to grow with us. It will encourage loyalty for them to achieve superior career growth in this company.

Therefore, internal communication between boss and subordinates must be in great terms. We need to keep motivating our staff. Even my children, they had to undergo normal working hours and policy. The similar working process and system apply to every staff under LBS as I expect this organisation to be professionally run with an outstanding workforce.

14.26 Do You Mind to Share with Us Any Commandments to Be Learnt by Nascent Entrepreneurs Keen to Dabble in This Industry?

Once you wish to get involved in this industry, please give full commitment to the product and service. Housing delivery should entail reasonable cost with good quality. Being pragmatic, I still need to keep learning consistently. One thing for sure is that as you grow older, do try to forgive people surrounding you. Learn to forgive and forget. Do not scold other people as it will trigger negative perception and other harmful situations. Do not argue too much and try to seek right solution for the benefits of mutual parties.

Chapter 15
Ken Holdings Berhad

Keywords Ken Holdings Berhad • Sam Tan • Green development • Sustainability • Scrupulous developer • Japanese working ethics

Today, Ken Holdings Berhad has emerged as one of Malaysia's elite green developers since they are recognised for their enduring dedication to raising the green profile of the nation. With an overall strength of less than 200 staff, this public-listed company is considered small in size. Notwithstanding the disadvantage in terms of working force, Ken Holdings Berhad has a strong engineering background with a 35-year history in Malaysia as it started out as a specialist engineering service provider and has received ISO certification in its respective sectors. They are now working on the projects in other states, mainly Kelantan and Johor, and it is their aim to apply their green building philosophies to these developments. Their products are easily fully booking, usually completely sold within 1 month. With an owner occupancy usage of almost 99%, Ken Holdings Berhad continues to enjoy immense credibility as one of the top-notch green developers of apartments and homes in Malaysia.

Historically, Ken Holdings Berhad was established by Datuk Kenny Tan. This property magnate is a relentless sort of a guy who does not give up easily and has a passion for the environment as well as for educating kids to build sustainable and high-quality communities. That has been the backbone to all his property projects and quickly making their mark in the competitive property development market. His great insight and intelligence allowed him to incorporate pre-existing concepts and turn it into ideas that have swept the world. Now, the tradition is emulated by his son, Mr. Sam Tan, who is currently the Group Managing Director of Ken Holdings Berhad. Mr. Sam Tan heads the property development, construction and sales and marketing divisions at KEN Holdings and is instrumental in spearheading the group towards green developments. Ironically, he is regarded as a 'green man' of Malaysia among fellow industry players. At the moment, his sisters, Jasmine and Yvonne, are

M. Jaafar et al., *Business Sustainability Model for Malaysian Housing Developers*, DOI 10.1007/978-981-10-5266-8_15

also working with the company. Like the family heritage, their property development skills have been well honed specifically on green innovation. Undeniably, Ken Holdings Berhad is the torch bearer of the green best practices and standards across the industry.

15.1 Please Narrate a Bit on the History of This Firm

This company was started in 1980 by our Group Executive Chairman who is also my father, Datuk Kenny Tan. He started as an engineering contractor. We were doing a lot of infrastructure works such as highways, buildings, bridges as well as land reclamation, and gradually we moved into property sector. Prior to that, he was involved in trading with the Japanese that had influenced his mindset, values and principles in business. I would say that my father is a business-minded person, and I think what he was specialised in was to bring and learn new technologies in Malaysia. That was his skills and forte. Back then, he had pioneered soil nailing in Malaysia, and he was the one who imported technologies here. He educated JKR (Public Works Department) engineers and how it is to be done.

15.2 How About Your Background? How Do You Get Involved in This Business?

I had obtained civil engineering degree in the University College London, United Kingdom, and pursued a Diploma in Law in the College of Law. After working as a civil engineer in London for three and a half years, I came back to Malaysia and work with my father in this firm. Being exposed to the property business at a young age had planted the desire in me to build homes for people. Obviously, I joined Ken Holdings Berhad in 2004. I took civil engineering first before pursuing my law which I had interest on, and I think it helps me a lot in business especially in analysing things, decision-making and what we can do or not. It is basically an additional learning for me. Anyway, I have been preparing myself my whole life to do this since I was in England.

15.3 What Potentials Do You See of This Firm During Your First Year Here?

As for property developers, the gestation period takes longer time. You need to purchase a land, apply conversion and get approvals as well. Also, you must see the changes in the market and what are the needs of the people, and these things were implemented when we first started the development of our condominium project in Petaling Jaya in 2000. When I first joined this company, it has already been doing well and got its name in the development field with a project named KEN Damansara. Just at the time I came back, we have some projects underway such as KEN Damansara 2 and 3 and KEN Aman in Seri Kembangan, and we were in the midst of thinking the plan for KEN Rimba.

15.4 When Did You Take over This Company from Your Father?

It was in early 2013, and it only entailed the change of job title as my father is still the key person in this firm. This situation is anticipated since the group is growing as we are delving into hospitality, building leasing, retail management and many other different businesses that insist us to do so. Still, property and construction play the big part of this company. In order to create distinctive personality, we are focusing on green development as our expertise. Even before 'green' became the very fashionable word, we have already implemented it in the earlier days. When we first started our high-rise project, my father had already perceived on the way to make the building cooler for the sake of its occupants and what materials need to be utilised for the homes to become cooler. Then, for our second project, we created natural ventilations through the corridor. All these things were planned in 1997/1998 regarding the application that involves generic innovative value and principles. For us, we started the green application not only to get the certificate, but more importantly, we create the right things for the right product or people. Also, as a homebuyer, you will get comfortable with your home, and the house value gets appreciated, and that is a very important aspect for us.

15.5 Please Share with Us When Was the Firm Able to Foresee the Opportunity of Green Technology

As the eldest son, I was blessed with the opportunity to see my father at his workplace. By 2005, Singapore had started its BCA Green Mark Scheme, and that year was the primary moment when people started to talk about sustainability. Paradoxically, we have already practised the principle of green sustainability while people had just talked about that term. Few years later, our project, KEN Bangsar, has been certified and recognised as the highest national achievement in green development. Ken Bangsar is the first project being awarded BCA Green Mark Gold^{PLUS} Award by the Singapore Building and Construction Authority outside of Singapore. I believe all of the achievements come from the engineering thinking and background. Whenever we decided to build, we must know how to build it first and identify its effectiveness. 'How are we going to reduce the waste?' and 'how to make it faster with better quality?' are the questions that have been embedded in our mind.

15.6 I Would Say that Your Father Is Your Mentor, so What Have You Learnt from Your Father?

How the company was managed from day one has inspired me to continue sustaining this business. As a very prudent entrepreneur, my father also is quite conservative as he will make sure that whatever we do will look well. We need to handcraft everything and put a lot of efforts into the work that we do.

But what drives us today actually resides in five simple but meaningful core values. The first is to be hands-on and committed. We are very hands-on in the decisions that we make, in the selection of materials, ideas and concepts, and very committed to what we do. Secondly, everyone must take pride in their work. If you are not proud of what you do, you would not do it well. It will be shown in the final results of your product. The third one is through the discovery of new ways to do things to create value for people. We give buyers something more than the charged price, so, as a result, customers will tell others about our good service and product. Another point is we believe that all team members are part of our family that play a critical role. Once they take care of family, then the family will take care of them. You need the whole team to pull the business success. And finally, *Mottainai*, which means too precious to waste in Japanese, that can be applied in every sense, in terms of resource, expediency in doing work and energy and in terms of time as everything is too precious to waste. What we mean by sustainability lies logically on our perspective that we can still live in the same comfort based on what we used to, without using as much resources as possible. Here, we can illustrate through the practice of Japanese that when they eat fish, they eat everything from the eyes to the scales and the fins, except the bones, which they will grind and turn into fertiliser. Hence, not a single part of the fish is wasted. In reality, this is what the company is exercising either physically, mentally or spiritually. Every staff knows about our inspiration on this matter.

15.7 In Terms of Personality of Your Father, What Are the Traits You Envision that Lead Him to Success?

Similar with the said core values, he is very hands-on and committed towards his work. Apart from that, he is keen on details, very particular, that you might not even see. Those are the characteristics that I am learning and trying to imitate. Even now, it is part of the company's culture. Do not waste and try to recycle or reuse because once you start wasting, it would probably be leakages in certain areas. Other than that, his determination to stay focused and innovative has aspired me to become an expert in this industry, by concentrating on green development and ISO administration.

15.8 How Does He Manage This Company?

He stresses on business strategy. He envisions that to keep growing, we need to implement new business acquisition and envisage bigger picture of corporate. We need to know the exact path that we choose. By way of illustration, we are planning to go for hotel ownership and engage external parties to diversify and expand our business. Also, he always alludes to the concept of teamwork among the staff for us to acquire vivid picture in the lens of organisational viewpoints. Working in team will enhance the operational strategy. That is where we segregate things through clear order and job responsibility as everyone looks after on each portion, respectively.

15.9 With the Sluggish Movement of Housing Market, What Are Your Strategies to Tackle It?

We believe that if you develop good quality of properties, people will always buy and recommend your products to others. In Malaysia, the buyers are ubiquitous. In order to keep up with the pace, we emphasise on the ISO practice in management field especially through quality of materials and working culture. The demands are always there due to growing population, and our majority population is young. This young population will build own family and needs more spaces for shelters. Notwithstanding the high demand from the growing population, the exogenous factors seem to deprive us to perform better. Some problems on bureaucracy vagaries and loan approval for potential buyers have inherently denied our profitability since banks are getting stricter and always associate the disapproval reasons with too much commitments or income barriers.

I think if the government continues to restrict supply or increase the cost of doing business, private housing developers will eventually feel the consequences. As a result, the supply will be reduced, and the economy will also be affected although Malaysia is actually healthy in terms of population with steady birth rate. In light of such circumstances, Ken Holdings Berhad is still capable and has been successful in understanding the emerging property market trends by incorporating novel ideas into the latest projects.

15.10 How Do You Acquire Your Land Banks and Where Is the Target Area?

Normally, we either buy the lands or joint venture with the state corporations to develop the housing projects. We are anticipating to cover all over the country with our green concept. Currently, we have projects in Selangor, Kuala Lumpur, Perak, Pahang, Kelantan and Johor. Our aim is to plant a green flag in every state in Malaysia. Similar to other large developers, we will also try to reap the benefits of national hot spots such as Penang, Klang Valley and Iskandar Malaysia. We receive remarkable invitations from every state for us to assist on the implementation of green development. But it is quite time-consuming to adhere on such needs because the invitations came from different backgrounds, and it is hard for us to accommodate their needs.

Our reputable name is getting well known among Malaysians as we are currently in collaboration with the state-owned enterprises of Pahang and Kelantan to build the green development there. But perhaps, the plan will be on mixed developments that comprise commercial and residential buildings. We are looking to cater the niche market, while it will be good for us to gain experience and create new lucrative investments since both locations offer interesting market to be explored. For your record, in Pahang, we will be joint venturing with the state for a project in Genting Highlands, while in Kelantan, the state government is envisaging to develop a conjoined block of hotel and serviced suite to supply the needs of local professionals and expatriates. Both projects are initiated based on good partnership and opportunities. We believe that the developers are the nation's builders, who may implicate national economy since their presence is significant.

15.11 Nowadays, Many Developers Refuse to Have Many Branches Because They Do Not Want to Deal with Different Local Authorities. So, How Do You Cope with This Circumstance?

We are lucky because the local governments are very receptive as they were also encouraged to have green development at their places. I would admit that every state differs in their land policy. We have a quite large of land bank that we believe useful for our future projects. We do not move everything at one time as we apply one project basis for every projected period through step-by-step approach, in order to ensure the project's success.

15.12 What Is the Annually Target for This Company?

Frankly, there are no specific financial targets; it may be based on personal achievement, just to maintain our performances slightly better from the previous years. We are projecting to have steadily growth every year. Nevertheless, the most important criterion is sustainable and recurring income. Thus, we intend to apply the said idea to our new head office, named Menara KEN @ TTDI to obtain consistent income. Besides, we are looking forward to have rental units, investment on buildings and hotel management as long-term strategy to sustain the business profitability.

15.13 How Do You Manage Your Staff? What Are the Mechanisms Used to Ensure They Will Be Loyal to Your Organisation?

We look at the attitude first in selecting new staff. If you have the right attitude, we will groom you to be better. If they have qualities, believe with our core values and take pride on their works, they surely can perform and be loyal to us. We have seen many proofs of individuals who started as junior executives and, at the end, becoming the leaders in the company. You must be committed and proud on what you do. If you are doing this for monthly pay, then other firms can also offer the same environment because we want to form a good team that treats us like a family. Of course, we keep improving in terms of giving benefits to our staff.

In terms of sustainability, we encourage our staff to use hybrid cars. If they buy hybrid car, we will pay the interest for the car loan. Also, we reward our excellent staff with point collection. From the assembled points, they can redeem into other benefits like medical prescription or buy glasses and contact lens, being subsidised by the company. We have holiday homes in Fraser's Hill, we charge a very reasonable and cheap rate for our staff, at least to cover the cost only. Aside from that, we offer special discounts to our staff to buy our properties. In fact, there is also an award for the best colleague, voted by their peers. I admit that we are still improving our human resources policy in order to create conducive environment for them to enjoy their work.

In my opinion, we encourage our staff to explore their talents. We extremely give chance for extra learning and guide our employees by putting them in different fields. This pragmatic approach seems viable to be implemented since it generates interest on job portfolios. In spite of that, I believe on the importance of staff training. We are eager to send them for training and seminar to understand the core values of business in order to accept easily the orders from management. We plan also to let them participate in courses that highlight the entrepreneurial features specifically for managers and head of departments.

15.14 Do You Have Any Management Structure to Gauge the Staff Performance?

We invent a new system that enables strategic alignment meeting to be conducted weekly. Managers and leaders will attend the meeting in order to ensure that the target is met and well-aligned. They set the key performance indicators for every quarter and monitor the schedule every week. It is pertinent for them to voice out every single thing, so that every department can work together to overcome the arising problems.

15.15 How Do You See Housing Development as a Risky Business?

I think it is a very long-term business. We are aiming future achievements with a long-term strategy. At the moment, the company operates in net cash position with zero gearing. We buy land in cash, but still, we have to be prudent in expense and seek for recurring income to continuously generate money. The graft for our share price is steady over a long period of time. It is steadily increasing, but again, we are still watchful against any unpredicted external force through high surveillance of supervision. On a separate note, the worst thing that worries buyers is related to abandoned projects. We always pride ourselves to deliver early and ahead of schedule. That is important to us.

15.16 How Far that the Use of ICT in Your Company Made Contribution to Success?

We update our website regularly, especially when we have new developments to be promoted. We use a lot of e-marketing through Facebook, Twitter, Instagram, etc. In particular, it is more to sharing pictures and lifestyle if we utilise the Twitter application only. Some tools work better than others. Different channels cater and provide different awareness. Above all, we try to improve on that matter, perhaps explicitly on the capability of our staff in embracing the use of ICT for the sake of work.

15.17 Apart from e-Marketing, Do You Implement Any Other Strategies to Market Your Product?

On top of e-marketing, we strongly believe that the role of word of mouth is very powerful and effective to promote our products. It is interesting to note here that we receive a number of calls and enquiries regarding the projects that had not been finalised yet. I believe it comes from the friends' recommendation. We used to engage real estate agents before, but now we have sales administrator that works under multitasking atmosphere, not only handles the sales and marketing but to handle strata title issue for administrative purpose. At this time, we use in-house and recruit our own sales team to close the deal by ourselves.

15.18 Which Category of Houses Are You Focusing on? What Factors that You Think Made Your Products Preferable?

We give a very good house yet made it affordable. We do not want people to misunderstand that with green application, it must be expensive. We used to sell terrace houses in Shah Alam for RM380,000/unit only. Today, the transaction is appreciated to RM650,000. So, the value is almost double. We are concentrating on houses that offer new concept of green. For instance, our new condominium in KEN Rimba, the price is started at RM399,000 – 1119 square feet with three bedrooms and two bathrooms. We predict that it will be suitable for young professionals and new family, with combining income of RM5000.00. Thus, with such privileges, it will create confidence among the buyers and subsequently enhance our brand. We deem that branding is the chief facet for product saleability. You may notice that all our properties are using KEN in their name. It is very risky at first but due to our confidence and commitment, we dare to tell the public that we are from KEN Holdings Berhad. That is an honour or pride to the company. I think we are the pioneer to use this strategy for the public to easily remember our brand. Apart from that, we try to maintain our relationship with the buyers as our target is to have at least 1% of repeat buyers. We are proud to say that we always exceed the figure. After-sale service is practised in order to please the buyers. We constantly attend the meetings organised by the Joint Management Body (JMB) to build trust and mutual bond with our customers. With this approach, we can tackle any arising problems and handle the buyers easily.

15.19 Could You Please Tell Us the Profit Range of Your Business?

People will always ask us how much is the cost for green application. Well, the costs and profits are depending on the project. The cost of green technology for KEN Rimba will go up to 5%, and we do not mind as long as it brings value and good picture to the buyers. We put the latest engineering technology in our product to make it look more trendy. Our profits are not very high. We possess our own construction arm, and both development and contracting firms are efficient and deliver good outcomes too. For us, the most important thing is the final product from the synergy of both arms. We need to give very good price to people, so that the houses will sell well. It is appealing to take note that we provide no brochures for our developments as people have put their trust on us. During any property fairs, we only bring the master plans and housing models; otherwise, they may come to our show houses to have a look.

15.20 Who Are the Buyers of Your Housing Products?

We have a lot of repeat buyers. Many of them are local buyers. It is our pivotal strategy for the buyers to grab our products as early as possible since due to great demand, we are capable to increase the price for the following phases. For KEN Rimba, we have buyers from outside Kuala Lumpur like Sabah, Melaka and Johor, and it seems that our plan has been fruitful. They registered with us through our website. They look for properties that have good investment in the future as most of our products are appreciating or experiencing price double within few years.

15.21 Any Other Salient Factors that You Think Might Be Important for Housing Developers to Consider? Have Economic Factors Influence on the Profit and Selling Trend?

Location is important and the genesis is based on the selection of land bank. We manage to get good land because we pay cash. Housing developers need to be hands-on and capable to study the site location. In my opinion, everybody will be affected with the macroeconomic movements. But we believe on our branding since we sell something that is different in concept. Take the case of KEN Rimba; we called it 'green cage' that encompasses green facilities with nature ventilation and lights, and it inextricably offers different concept, furnished with good landscape of jogging track, special side lane and others of the ilk.

15.22 Do You Plan to Diversify into Other Industries?

Now firmly in property development, Ken Holdings Berhad is spearheading green building developments around the greater Klang Valley area with modern design and ingenious facilities at very attractive locations. Other than that, we seriously plan to have the recurring income through our properties which commensurate with the sustainability model. This idea has actually inspired the company to look at the hospitality sector as well. Our land in Penang and Genting Highlands is deemed suitable for hotel development. Thus, hospitality is under the pipeline. Future projects may include the plan to actively participate in the development of Kota Bharu, Pahang, Penang, Melaka and Johor Bahru although land is scarce especially in prime area.

15.23 Anyway, Are You Active in Any Corporate Associations?

At REHDA, I get involved in elevating the reputation of developers in Malaysia. I have successfully launched the Green Tour Series and a Green Rating and Carbon Calculator Tool called GreenRE.

15.24 We Believe that with Such Novelties, This Firm Must Have Collected a Number of Awards. May You Please List Out the Awards Received by Your Esteemed Organisation?

We are the first developer to develop green building in Malaysia and also the first Malaysian company to consistently achieve carbon neutral status since 2010. The first award was in 2009; since then many awards came by with KEN Rimba contributing a lot. For a record, KEN Bangsar is the first Malaysian project to receive BCA Green Mark Gold^{PLUS} Award in 2009. Ken Bangsar has also won the Green Building Index Gold Award, the Building and Construction Authority (BCA) Green Mark Gold^{PLUS} Award and the FIABCI Malaysia Property Sustainable Development Award in 2011. Besides that, KEN Rimba is regarded as the Malaysia's first multiple award-winning green township. The most recent one is the FIABCI Malaysia Property Sustainable Development Award in 2014 won by KEN Rimba Legian Residences. Our upcoming corporate office, Menara KEN @ TTDI also obtained multiple platinum through BCA Green Mark Platinum Award (Provisional) 2013 and the USGBC LEED Platinum Award (Pre-Certification) 2013.

The following list contains the record of accreditations and awards received by us:

Project	Awards/recognitions
KEN Bangsar	2009: BCA Green Mark Gold^{PLUS} Award
	2010: GBI Gold Award
	2010: The EDGE-PAM Green Excellence Award
	2011: FUTURARC Green Leadership Citation Award
	2011: PAM Silver Award for Excellence in Architecture
	2011: FIABCI-Malaysia Property Award – Sustainable Development
KEN Rimba Shah Alam	2011: GBI Pilot Project for The GBI Township Tool
KEN Rimba Legian Residences	2010: BCA Green Mark Gold Award
	2010: GBI Certified Award
	2014: FIABCI-Malaysia Property Award – Sustainable Development
KEN Rimba Commercial Centre	2010: BCA Green Mark Certified Award
KEN Rimba Jimbaran Residences	2012: BCA Green Mark Gold^{PLUS} Award (Provisional)
	2014: GreenRE Platinum Award (Provisional)
KEN Rimba Condominium 1	2013: BCA Green Mark Gold^{PLUS} Award (Provisional)
	2014: GreenRE Gold Award (Provisional)
KEN Rimba Condominium 2	2013: BCA Green Mark Platinum Award (Provisional)
	2014: GreenRE Platinum Award (Provisional)
Menara KEN @ TTDI	2013: USGBC LEED Platinum Award (Pre-Certification)
	2013: BCA Green Mark Platinum Award (Provisional)

15.25 If You Do Not Mind, Please Render Some Advice for Upcoming Housing Developers

Our business is based on reputation and it takes many years to build. Do not enter this industry only to make quick bucks. You have to be accountable with your works by supplying quality houses to people for national building. Take the responsibility seriously to deliver the best of your efforts.

Chapter 16
Mah Sing Group Bhd

Keywords Mah Sing Group Bhd • Leong Hoy Kum • Integrated developer • Value creation • Customer care • International accolades

In property development, Mah Sing has progressed by leaps and bounds since its inception in 1994. Today, having carved a name for itself in the property industry, it is a well-known and much-respected property company with unrivalled stature especially in the residential and commercial segment. Tan Sri Dato' Sri Leong Hoy Kum had built the company from the plastics trading firm his father had established in 1965 to the RM3 billion corporation it is today. He basically built his company from scratch and is now the largest non-GLC property developer, with market capitalisation of RM3.61 billion. Not only that, Mah Sing Group Bhd has become a notable housing empire by winning numerous accolades from both national and international recognitions. To date, the group has 46 projects spread throughout the Klang Valley, Johor Bahru, Penang and Kota Kinabalu.

As the largest entrepreneur-run property developer in Malaysia, Mah Sing is often in the headlines with its land acquisitions and innovative projects. Being at the forefront of this very demanding segment, Mah Sing has unfailingly delivered success after success, most times surpassing industry standards and benchmarks. Delivery beyond comparison is almost synonymous to the Mah Sing brand due to dynamic team of experienced and dedicated property professionals working together in synchrony. Having undergone a corporate transformation in the early stage of 2017, Mah Sing is one of the leading organisations in Malaysia to champion the green agenda, building sustainably and maintaining a focus on viable alternatives. Befitting with their new tagline 'Reinvent Spaces. Enhancing Lives', the group is well known for its urban development features that play a role in contributing to a better, healthier lifestyle and never fails to use innovative construction materials as well as incorporating environmentally friendly facilities in its townships. These efforts have garnered the group with the esteemed Malaysia Green Tech Developer Award in 2013, Property Insight's Best Eco-Urban Township 2016 for Southville City and a total of 12 Green Registrations which include Malaysia's Green Building Index, REHDA's GreenRE, Singapore's Green Mark and USA's LEED. Besides

M. Jaafar et al., *Business Sustainability Model for Malaysian Housing Developers*, DOI 10.1007/978-981-10-5266-8_16

being the man at the helm of Mah Sing, Tan Sri Dato' Sri Leong Hoy Kum is also on the central committee of the 900-member Malaysian Plastics Manufacturers Association. He has been the Honorary President of the Young Malaysian Movement Association (YMM) since 1999. Further to this, he sits on the Board of Directors of various other private companies. He was among the 27 winners of the Asia Pacific Entrepreneurship Awards 2013 for 'Entrepreneurs of the Year'. In recognition of his achievements, he was conferred the Darjah Paduka Mahkota Selangor (D.P.M.S.) which carries the title of Dato' and the Jaksa Pengaman (J.P.) awards by the Sultan of Selangor in 1996 and 2001, respectively. In 2007, Tan Sri Leong was conferred the 'Darjah Kebesaran Sultan Ahmad Shah Pahang Yang Amat DiMulia – Peringkat Pertama Sri Sultan Ahmad Shah Pahang' which carries the title Dato' Seri, followed by the 'Darjah Panglima Setia Mahkota' which carries the title Tan Sri in 2009. More recently, Tan Sri Dato' Sri Leong has acclaimed property division's most prestigious personal accolades such as the BrandLaureate Special Edition World Award in 2015, Asian Corporate Director of the Year 2015, the Edge Most Outstanding Entrepreneur 2016 and Sin Chew Business Excellence Award Lifetime Excellence Achievement 2017. After more than two decades in the industry, he is still going strong. His business verve is unabated, but he is under no illusion that this will continue and is infusing new blood into the business. Presently, his daughter, Jane, has been elected to join him as one of the board members.

16.1 Could You Please Tell Us a Bit on Your Background?

I was born in Alor Setar and my parents moved to Kuala Lumpur (KL) at a very young age. Nevertheless, I consider myself a KL boy. My family lived near Jalan Ipoh where I studied at a Chinese school before going on to Methodist Boys' School in Sentul, alumni of which include Tan Sri Azman Hashim of AmBank Group and ex-MCA[7] President, Tan Koon Swan. My father founded a plastics trading business back in the 1960s, but I initially did not want to join the family business as I wanted to be an engineer. But being the only son, I was eventually persuaded to take up plastics training in Japan, and upon my return to Malaysia, I grew the business into manufacturing of proprietary and OEM plastic products and listed the company on Bursa Malaysia in 1992. I was involved in all aspects of the business, from product design to marketing and sales to quality control and just-in-time delivery of the products to my customers. In 1994, we were facing thinning margins, and in a bid to protect my shareholders' value, I knew I had to forge a new path for Mah Sing. At that point of time, I had a very strong interest in property investment, and I felt that this vibrant sector could provide the next impetus of growth for Mah Sing.

Thus, in 1994, I entered the property development arena with Mah Sing's maiden project in Ulu Yam, Selangor. The 45-acre project with a gross development value

(GDV) of RM21.5 million featured link houses. Once I started, I got hooked. Besides a new-found passion for real estate, I also saw the property industry as an avenue to maximise shareholder value. When I cut my teeth on property development, I had to separate the highly successful and established Mah Sing Plastics from the new entity of Mah Sing Properties. This approach ensured that both grew organically and did not compete for my time and resources. I would consider myself as a fighter and a very ambitious businessman. I always set a high expectation for myself. I want to transform Mah Sing into a global property developer, a great, lasting and profitable company. Today, Mah Sing is one of the few fully integrated property developers with high-rise and landed residential as well as integrated commercial and niche industrial parks.

16.2 Tell Us How You Came to the Decision to Venture into Property Development? What Inspired You or Was It a Purely Business Decision?

Mah Sing Group Berhad started out as a plastics trading house, went on to be listed on the Second Board of the Kuala Lumpur Stock Exchange in 1992 and was redesignated to the property sector in 2000 culminating in its transfer to the Main Board of Bursa Malaysia Securities in 2004. I have always had the vision to grow the company into one of the top in the business. After establishing Mah Sing as an industry leader in the plastics business, I looked for ways and means to increase and enhance the shareholders' value. Through my satisfaction in personal investments and passion in designing my own home, I decided to venture into property development in 1994. At that time, the property sector was going through an up cycle, and I saw the potential of improving shareholders' value through property development. The company emphasises on maximising returns to the shareholders, and it is through this philosophy that Mah Sing has grown from strength to strength since 1994.

The initial years were learning years for us and our capital base was small. We have performed especially well in the past 7 years where our revenue rose by more than 2-folds to RM701.6 million and net profit grew nearly 18-folds to RM94.3 million up to 31 December 2009. This is a healthy compounded annual net profit growth rate of 51%. Anyway, for your record, in 2013, our revenue was RM2.01 billion, with profit before tax of RM371.50 million. Meanwhile, in 2014, our revenue rose to RM2.90 billion, with profit before tax of RM455.01 million. 2015 marked a year of growth once more as it shows Mah Sing is heading in the right direction with revenues hitting the RM3 billion mark (RM3.1 billion), with a profit before tax of RM386.67 million.

16.3 How About Your Plastic Manufacturing Business? Do You Mind to Share Its Current Business Growth?

Today, the plastics division is going strong with operations in Malaysia and Indonesia. There is a separate management team for the plastics division, and as they are very capable, they have been running the division independently for many years. The plastics division has been steadily winning various corporate and product awards, and we have our own patent-pending products. The Malaysian operation is a wholly-owned subsidiary of the group, while the Indonesian operation is a 65%-owned subsidiary.

16.4 How Do You Define Success in This Housing Business?

A successful developer should be able to offer right type of products to fit the property cycle. They should also have adequate funding and support from the shareholders and bankers as well as continued government support. Nevertheless, a successful developer should be able to deliver what it promised and be able to provide strong customer services.

It has taken us years to be what we are today and our success is a collective effort. It may sound simple, but a lot of research works and commitments are needed to ensure that the product fits buyers' needs and aspirations, at a relevant price point. An important lesson learnt is to continuously strive for better results. We must always remind ourselves to be humble so that we do not get carried away during our voyage. This is how we have grown from strength to strength, and our balance sheet has stayed very healthy for further expansion. A successful property developer should aspire to create value in addition to building quality homes.

16.5 What Are the Factors That Contribute to Mah Sing's Success?

I would attribute our achievement to the right business model, dedicated staff, good and experienced pool of contractors, architects, consultants and suppliers and last but not least our supportive customers. Our key success factor is our business model for the past 10 years which hinged on fast turnaround. Land acquisition in multiple locations and quick launches allowed us to tap immediate market demand and enjoy good take-up rates. We invest in research and development to ensure the projects meet buyers' needs and complement their lifestyles. Timely delivery is also critical and we are proud of our good track record. We have extended our strategy in recent years to reflect our current scale of operations. We are ready to acquire and develop large tracts of strategic land and look to venture overseas for new sources of earnings growth in the longer term, but the core of speed, discipline and innovation remains the same.

Of course, nobody can do all this alone, and I attribute our success today to our great team who are dedicated, skilled and loyal professionals. I can rely on them to execute what has been planned. Trust is very important and my management team knows that I trust them to do their job. When everyone is efficient, the whole organisation benefits, and people get to spend more quality time at home as well. Over the years, we have also cultivated a panel of trusted contractors. We engage good and experienced contractors, architects, consultants and other suppliers who can deliver our vision. We also have a strong base of customers, and we recently launched our reward programme, the M Club. Among the privileges to be enjoyed by members are repeat purchase discount, buyer-get-buyer rewards as well as birthday surprises with more rewards to be introduced in the near future.

16.6 Do You Think that Your Prior Related Experience Contributes Towards Business Success?

My training in Japan had taught me to emphasise on quality, cost-efficiency and on-time delivery and always deliver the best service. As a manufacturer in my early days, I learnt that control, cost and fast turnaround from raw materials to finished products are very important to survive. That is how I continue to ensure that these controls are incorporated into the property development business. I also learnt good planning from the Japanese. They may take a bit longer and are very cautious, but once it comes to execution time, their efficiency is second to none, consistent and positive. I believe in extensive analysis and prudent evaluation and taking calculated risks in life and business. If I do not take risks, I would never gain as the Chinese saying goes, 'in danger lies opportunities'.

16.7 May You Share with Us How Did You Manage to Succeed and Overcome the Challenges from the Beginning of Your Venture in This Property Development Industry?

We had a steep learning curve in the beginning. When we first started, we did not have a strong balance sheet, nor a large land bank. To overcome this, we adopted a quick turnaround strategy which helped us to be capital efficient as we rolled over our capital and balance sheet, and this helped us to grow our balance sheet quite quickly to take on larger projects. With a stronger financial standing, banks were willing to give us a more favourable borrowing rate compared to when we first started. We made a conscious decision to go medium, medium-high and high-end with our properties, and our mantra was 'location, quality and branding'. The Mah Sing brand encapsulates the concept, product quality, standard of design, customer service and track record. We package all that with the right location and strategic marketing.

We have also moved from Johor Bahru to Klang Valley (Greater KL) and now Penang Island as well as Kota Kinabalu, Sabah, as we wanted to establish ourselves in the four property hot spots in Malaysia. Most of the property transactions take place in these locations and we want to capture the market share. Another challenge that the company faced as a developer was during the Asian Financial Crisis of 1997. At that time, consumer sentiment was low; however, Mah Sing was still profitable because of our quick turnaround business model which has been the backbone of our success over the past decades.

16.8 In Your Opinion, What Are the Traits That Should Be Possessed by Businessperson Like You in Order to Succeed in This Industry? Any Business Mentors That Have Inspired You?

A successful businessman is someone with a vision, can take calculated risks and has the tenacity to go the distance. I admire people who possess great wisdom. Wise people are those who know what to do and when to do it. This means the person is able to consider all angles of a situation and is able to act accordingly and respond to the situation in a timely fashion.

My father was the first wise person that I look up to. It was his pioneering spirit, the way he saw opportunities for growth before they were even there, that really inspired me. When I look back now, if it wasn't for his constant encouragement to consider plastics training, I might not be where I am today. Also, I have admired some successful entrepreneurs like Warren Buffett and Li Ka-shing. Their perseverance and hard work coupled with their sharp business acumen all work together to their success.

16.9 Please Share Your Business Principles that You Apply to Sustain High Achievements

The Mah Sing brand has also been recognised as a much-sought-after property developer, and at last count, we have been awarded with over 100 local and international awards from various industry groups, media and the public for company performance, corporate governance, product design, concept, innovation and quality. Even though we have received numerous accolades, we still keep shooting for the stars and never give up; these have been, still are and will be the best advice ever given simply because, when you keep shooting for the stars, you keep aiming to be the best in your field and, when you do not give up, you persevere and push on despite trying circumstances. I keep shooting for the stars as I aim to be the best in the field, and it is easier to shoot when you have a target. Pursuing something with single-minded focus and prioritising items helps to drive success. In the meantime, never giving up means facing a challenge head on, breaking it down to manageable parts and getting the expertise of people familiar with the problem and resolving it.

16.10 What Are the Current Directions and Visions of Mah Sing?

We recently underwent a corporate transformation as a symbol of Mah Sing's identity, a company that strives for excellence in product delivery, quality, customer service and customer care. As we put on a new look, we rethink our purpose. Our tagline is '*Reinvent Spaces. Enhance Life*'. We want to be a meaningful corporation for now and the future. Now, in everything we do, we want to, firstly, enhance life. We want to ensure all live well: our customers, our people and our stakeholders. 'Reinvent Spaces' means that the Group aims to raise the bar of future living by recreating living spaces, work spaces, business spaces as well as social and community spaces. Hence, as our new vision states, we aim to invent a future living that enchances quality of life.

We have evolved from a developer building houses to building communities. For example, in Mah Sing's Southville City@KLSouth, we have a sustainable township with a cohesive master plan that delivers a conducive and safe environment. We have also adopted this model at our other townships such as M-Residence in Rawang and Meridin East in Johor Bahru. The competition among property developers in Malaysia is very high, and this is a good thing! It is through healthy competition that we can push ourselves to be better and to grow Mah Sing to where the company is today.

16.11 Which Department in Your Company Is the Most Critical and Plays Key Roles in Upholding Superior Performance?

I believe in human resource development to ensure loyal staff. They are the people who make the company. I am blessed to have a good and loyal team. I treasure members of every department and their experience and skill. They have immense control over the different parts of the business, be it business development, sales and marketing, project planning and management, CSR and other areas. We are a company that is highly invested in our human capital because we believe that our team is our greatest asset. Currently, we employ more than 800 people in our property division.

We are pleased that we are recognised as one of the Best Companies to Work For in Asia by HR Asia Awards. We were also awarded the Best Company for Leadership Award by IAIR Awards and the Best Managed Company in Malaysia by Euromoney Awards. More recently, we were awarded the Employer Branding Award for Malaysia Best Employer 2016. Team Mah Sing will further enhance the value of the company by providing great customer service and quality products and by innovating to meet the market demand.

16.12 Location-Wise, How Do You Implement Site Identification and Selection? May We Know the Total of Land Bank Owned by This Company?

As a market-driven developer, we go to where the demand is. Mah Sing's main focus continues to be in the four property hot spots where our branding has already been established, namely, Greater Kuala Lumpur and Klang Valley, Iskandar Malaysia and Johor Bahru as well as Penang and Sabah, with remaining GDV of RM27.73 billion for 2522 acres. These are also the high economic zones where there are high demands for property. Our portfolio of projects not only serves the upper end of the market but also provides an enhanced lifestyle for the masses. In fact, many of our properties meet the value-for-money demand of the middle-income group, who desire a higher quality of life in an affordable price range.

For example, government has identified Greater KL and Klang Valley as a National Key Economic Area (NKEA) under the Economic Transformation Programme (ETP) where it aims to grow the population from 6 to 10 million with a focus on growing foreign talent from 9% to 20% of the population as well as to create 553,000 jobs. We have identified that the demand for property is in this high population area. It is a matter of location, products and timing. Some of the lands that we acquire are in locations that are not traditionally viewed as prime areas, but we have done extensive market studies, and with infrastructure improvements, we have transformed the area.

By and large, we continue to scout for more land but we are not in a hurry. Usually, we reserve smaller pieces of land for niche products and bigger pieces of land for township development because we have the capacity to explore diversified product offerings. In our opinion, locations with excellent accessibility as well as potential for growth will continue to attract buyers and investors.

Our property mix consists of 33% of commercial projects and 63% of residential development while the remaining 4% are industrial. Currently, we have 35 active projects, all of which are in various stages of progress. Together with our existing projects, these developments will be built over the next 8 years. Mah Sing's six new projects are Cerrado serviced apartments in Southville City@KL South, Bangi, which are affordable luxury homes indicatively priced from RM388,000, new township Laman Ayu in Rawang, new blocks in D'Sara Sentral, Lakeville Residence in the central region, Ferringhi Residence 2 in Penang Island and Meridin East township in Pasir Gudang with affordable landed homes indicatively priced from RM350,000.

16.13 With Iskandar Malaysia Undergoing Massive Development, Will Mah Sing Acquire More Lands in Johor?

I am confident about the proposed Malaysia-Singapore Rapid Transit System (RTS) link which will likely connect Johor Bahru at JB Sentral with the Thomson MRT line at Woodlands North Station in 2019. This will improve access and property value and make it even more convenient for Singaporeans to move their investment to Iskandar or stay there. In that sense, we will continue to scout for more lands but we are not in a hurry.

I am very proud that Mah Sing has retained an entrepreneurial spirit and professional management. Building Mah Sing together with my team over the years, we have found that life is about continuous learning and improvement. The world is constantly changing. As a market-driven developer, we need to adapt to and stay ahead of the market. As we strive to live up to our brand promise, which is to be a premier lifestyle developer, we have learnt to invest strongly in our human capital. This is because, at the end of the day, it is my team who have led Mah Sing to where the company is today and it is my team who will lead Mah Sing to the next level.

16.14 For Future Plan, What Type of Houses that Mah Sing Will Look Forward to Be Developed?

Four to five years ago, most developers targeted higher-end properties. For the next 1–2 years, the trend is more towards mass market housing. Overall, 89% of our residential launches have been priced below the RM1 million threshold, with 68% being below RM700,000 and 50% priced below RM500,000. We will have that, the affordable range as well as the mid- to high-end. For example, our Garden Plaza suites in Cyberjaya are also being sold at a very affordable price, from RM318,800. While in Rawang, we have link houses priced below RM500,000.

We would like to continue our township development, niche industrial development as well as strategic commercial and mall development in tandem of enhancing our branding and after-sale services. We want buyers to know that we are no longer building houses but building communities. We may look into plans to diversify in the future, but for now, our focus will be on what we do best, which is property development.

16.15 Do your Company Utilise Any Marketing Strategies? Please Share Any Specific Tools That Have Been Implemented

Marketing is an important effort in gaining product attention, building the brand, servicing demand and winning customers. To me, marketing includes having the right products or services, selecting the right locations, implementing sales programmes and promoting the company to the public. That is why we always strive to offer the right type of products to fit the property cycle. We have a strong marketing team that ensure all our products are made known to the public through various communication vehicles such as traditional media as well as the contemporary media.

16.16 Why Do You Think the Buyers Will Purchase Your House? How Do Your Company Retain Your Buyers?

We always keep in mind that we want all our buyers to have good rental income and capital appreciation. We strongly believe in good after-sale service. Our customer service team is the first point of contact for buyers' feedback. We are always working to improve user experience at each and every contact point. It is up to us to plan how we want to attract more buyers. To do this, we will need to roll out products that are still relevant at a price point which people can accept. We will need to build products that are rightly priced.

We are optimistic of our strong sales momentum as our products cater to market needs and are located in strategic locations, without compromise on the quality of the product. We will also need to be able to provide the right type of product in different regions based on market demand in that area. Besides, we are always working to improve the experience of our users as they are our best brand advocates. With all these components in place, we can ensure that the quality of our products is of the highest standard and that we are able to continuously innovate to meet market demand and improve our branding. With 46 projects spread across Malaysia, we treat each project as a special opportunity to enhance the lifestyle of our buyers as well as enhance our branding. Be it a mass market product or a high-end one, each new launch allows us to raise the bar in developing quality lifestyle properties for our buyers. We are committed to deliver innovative concepts and high-quality and exceptional service and, ultimately, leave a legacy of excellence.

On top of that, customer satisfaction has been seen as a key performance indicator these days. However, we at Mah Sing aim to push that to the next level, which is customer delight. We wanted to give our consumer something that exceeds their expectation. Mah Sing has set up M Care to assist house buyers. M Care is a one-stop professional service for Mah Sing property owners to provide a myriad of services. The professional consultation services include renovation and interior design consultations, relocation services, butler or concierge services for owner-occupied units, as well as leasing and subsale services. M Club is a distinctive buyer loyalty programme specially created to provide additional benefits to house buyers where members can enjoy buyer-repeat purchase discounts, buyer-get-buyer rewards, birthday surprises and many more exceptional benefits. Currently, we have testimonials from more than 31,000 members.

16.17 Do You Set Any Future Plans for Mah Sing? Perhaps, Do You Have Any Upcoming Strategic Investments or Potential Diversification?

There are a few key principles when it comes to our planning and development. The key element is sustainability as we plan our projects to be self-sustaining and practise green building techniques. Our projects like Icon City in Petaling Jaya, Icon Residence in Mont' Kiara and many others have green certification like US's LEED, Malaysia's GBI and Singapore's Green Mark. Sustainability also extends to after our buyers collect their keys – besides being a dream home, their property ownership must make financial sense for them as well. We keep in mind that we want all our buyers to have good rental income and capital appreciation. Further to this, we want to have repeat buyers. Buyers today are very savvy, and they have a lot of choices. Ultimately we build for them, so customer service is very important to us as well.

16.18 What Are the Strides That Have Been Employed by Mah Sing to Give Back to the Society?

As a caring corporate citizen, Mah Sing's corporate social responsibility is deeply embedded in our corporate philosophy. Beyond making donations to worthy causes, we further set up a charitable trust, Mah Sing Foundation, to receive and administer funds solely for medical, educational and charitable purposes, as well as to seek to raise more funds via annual fundraising activities and events. It is indeed gratifying to find that our efforts have been appreciated. As an eminent figure in the society, I am very proud to receive the Asia HRD Awards 2014 for the contribution to society category. Aside from that, to date, the group have contributed over RM11.5 million to the social and economic development of the community, both locally and abroad, of which 200 contributions were for medical, educational and societal needs for the poor and underprivileged. We believe that an award such as this recognition, spurs improved performance of the award winner. As a market-driven developer, we not only build homes, offices and industrial lots, we also strive to build communities. This achievement is a mark of distinction on our social contribution activities, not just at the corporate and investment levels but also in the eyes of the public.

16.19 What Do You Think on the Role of Government in Facilitating Your Business?

We also laud the government's commitment to increase and improve road connectivity and the public transportation system, as this will directly and indirectly benefit our various projects which are in prime locations in Greater KL, Penang Island and Johor Bahru. While our projects already enjoy excellent accessibility, any additional connectivity will bring additional convenience for residents and tenants and has the potential to increase property values.

Aside from accessibility, we are heartened to see the government's continued commitment towards major projects such as Mass Rapid Transit (MRT) and Light Rapid Transit (LRT), Cyber City Centre, RAPID Pengerang and High Speed Rail (HSR). The government's continued efforts to make the High Speed Rail a reality and their commitment to improve highways, MRT and LRT will directly benefit Mah Sing as our focus is in Greater KL which makes up 60% of our land bank in terms of gross development value.

We have just recently launched our RUMAWIP (Federal Territories Affordable Homes Programme) project, Residensi Seri Wahyu of Taman Wahyu, in which we collaborated with the government to develop affordable housing for the public that is priced at RM300,000 and is targeted at first-time house buyers. While we carefully time our launches to ensure products are in line with market demand, we also actively pursue sales from existing projects. In fact, 89% of our planned residential launches for 2016 are below RM1 million per unit.

16.20 How Do You See the Importance of Economic Factors in Contributing Success?

We have learnt quite a few lessons during the last three economic crises. I would love to recall on our approaches during the 1997 Asian Financial Crisis. As a newbie, I was conscious that the first project could not fail as this might affect our confidence level. At that time, consumer sentiment was low; however, our group was still profitable because of our quick turnaround business model which has been the backbone of our success over the past decades. This was why we launched bread-and-butter products – single-storey link houses. With the strong sales, we were able to have a comfortable cash flow that tided us over.

Mah Sing has a strong team, a comprehensive range of products and a strong financial position. Hence, we believe we can weather challenging times better. We are careful to plan and maintain our market leadership position and to roll out products that are relevant at a price point based on market needs. We are a financially prudent company. We have kept our gearing very reasonable, and this will allow us to tap any good opportunities that may arise. We will also focus on offering affordable priced products as consumer spending power has been affected. At the same time, the quality must be there, and we will offer good concepts that will appeal to the younger generation. In today's time and age, quality is a given, and this is the reason we focus on delivering good quality with good customer service.

Currently, while market sentiment is poor, there is still a supply and demand gap – only 70,000 to 80,000 new homes are completed each year, while about 200,000 new properties are required to meet the number of new households formed from new marriages. This is why we still have selective launches. In the face of inflationary and economic concerns, we believe property will continue to be the preferred investment option to hedge against such concerns. We are practical in our product launches as we garner a decent 70% take-up rate for a project block before launching a new one. Suffice to say, in order to survive an economic crisis, I would say that financial stability, the right products and the right customers are crucial. There will be a niche segment in every economic cycle. We really have to understand what customers want and need.

16.21 Please List Down the Awards That Have Been Received by You and Mah Sing, Respectively

Since we ventured into property development, we have received over 100 international and domestic awards in recognition of our outstanding corporate performance and products. The most recent award might be from the prestigious 2016 Frost & Sullivan Excellence in Competitive Strategy Innovation and Leadership Award in Property Development that recognises companies for superior leadership, technological innovation, customer service and strategic product development within regional and global markets. Besides that, in 2014, I was inducted into the Asia Pacific Brand Foundation's Hall of Fame as I was conferred a 'Lifetime Achievement Brand Icon Leadership Award'. Anyway, the list of awards received in 2016 and 2015 are as follows:

2016

1. The 6th Asian Excellence Award – (Asia Best CEO)
2. Forbes Asia's 2016 Heroes of Philanthropy
3. International Entrepreneur Excellence Award 2016 by MACITA
4. Outstanding Entrepreneur Award – from MRCA Crown Awards Night
5. Lifetime Achievement Award by World Chinese Economic Summit
6. ASEAN Business Awards Malaysia 2016 – Entrepreneur of the Year 2016
7. Sin Chew Business Excellence Award – Lifetime Excellence Achievement Award

2015

1. IAIR Awards – Best Company for Leadership/Property Development/Malaysia
2. Property Insight Malaysia's Prestigious Developer Awards 2015 – Top 10 Developers

3. Social Media Excellence Awards 2015 – Lifestyle Developer
4. Malaysia Best Employer Brand Award 2015
5. The Edge Malaysia Outstanding Property Entrepreneur 2015 – Tan Sri Dato' Sri Leong Hoy Kum
6. Best Developer of the year in iPropert y People's Choice Award 2015
7. Corporate Governance Asia – Outstanding Companies on Corporate Governance
8. The Edge Malaysia's Top Property Developers Awards 2015 – Top 10 Property Developers Award
9. iProperty Developer of the Year 2015

16.22 Please Share with Us Any Specific Working Cultures of Your Esteemed Organisation. How Do You Treat and Manage Your Staff?

It is the spirit of camaraderie between departments and the sense of ownership, pride and accountability within the employees that will contribute to the business success. We are genuinely committed to attracting talents and developing and retaining our people, because human capital is the driving force of a successful company. We promote continuous learning and improvement of each individual and encourage work life balance. All employees are given the chance to improve via training programmes, workshops and interactive sessions. Henceforth, we provide strategic and continuous in-house and external training to employees of all levels. Employees can also partake in various activities outside of work such as weekly yoga classes, bowling tournaments, family days and community sports.

In line with our transformation, Mah Sing kicked off 2017 with the renovation of its headquarters in Wisma Mah Sing, Jalan Sungai Besi which will be known as MS@Work. The new headquarters feature various modern layouts that appeal to millennials such as open space concept, fun meeting rooms, collaborative breakout areas, colourful pantries, gyms, café and many more. We understand the importance of our staffs' well-being in the company. Our property division's workforce increased from 350 people in the year 2005 to 900 people to date. We have a high percentage of millennials in our workforce, and we believe that addressing their needs is important. We are transforming our company so that both new and experienced staff will be able to achieve a better personal and career growth. Apart from that, working environment is constantly updated to provide employees with the optimum tools to perform well their jobs at the workplace. Work-life balance is inculcated through various activities organised under MS Sports Club such as festivity gatherings and interdepartmental sports competition. We also sponsor employees to join community sports to foster competitive spirit, teamwork and camaraderie. Working environment at Mah Sing is continuously improved. Work stations are upgraded to provide employees with the tools they need to excel in their work. Valuing performance culture is practised to motivate employees to go the extra mile.

Family days with exciting activities are organised to encourage employees to spend quality time with family and friends. Various team buildings are also held to promote the spirit of teamwork.

16.23 Lastly, Could You Please Provide Some Wisdom Words to Be Pondered by Fellow Players?

One has to be a good thinker and planner in offering the right products for the market through good quality control, market research, product development, branding, sales and marketing and customer service. It has taken us years to be what we are today and our success is a collective effort. It may sound simple, but a lot of research studies and commitments are needed to ensure that the product fits buyers' needs and aspirations, at a price tag that they can afford. An important lesson learnt is to continuously strive for better results. We must always remind ourselves to be humble so that we do not get carried away during our voyage. This is how we have grown from strength to strength, and our balance sheet has remained very healthy for further expansion. Essentially, the secret to be a successful developer is to anticipate and give buyers what they want, with a value proposition that they can appreciate.

Chapter 17
Titijaya Land Berhad

Keywords Titijaya Land Bhd • Lim Soon Peng • Family business • Contractor • Humanitarian • Project location

Titijaya had its beginning way back in 1997 in the property development and construction industry. It was founded by Tan Sri Dato' Lim Soon Peng, a visionary businessman in property development. Since its formation, Titijaya has embarked on numerous development projects, from double-storey terrace houses and three-storey shop offices in Taman Bukit Cheras, Kuala Lumpur, to residential lots and low-cost apartments in Klang, Selangor. Tan Sri Dato' Lim Soon Peng, the Group Managing Director of Titijaya, has accumulated more than 40 years of experience in the property development industry and is responsible for the overall business strategies of Titijaya. Under his leadership, Titijaya has grown into one of the most established property developers in the Klang Valley. Titijaya is recognised for its innovative signature properties with award-winning and attractive lifestyle designs.

From a small player in the market, Titijaya has become one of the prime property groups in the industry and is well known for its quality projects that have been built and handed over ahead of schedule, enabling their customers to gain faster capital appreciation as well as savings. Titijaya's core business is focused on three segments, namely, residential, commercial and industrial property developments. Today, Titijaya has a portfolio of completed development projects with a total gross development value exceeding RM2 billion. Their portfolio now includes duplex living, office suites and towers, housing communities and apartments, business centres, hotels, shops and mixed development projects. Titijaya Land Berhad was listed on the Main Market of Bursa Malaysia on 27 November 2013.

Together with a far-sighted vision, i.e. 'Aspire To Be The Best, Grow Rapidly, Mould An Excellent Team And Winning Culture' and backed by a passionate mission, i.e. 'To Build Properties That People Will Buy, Appreciate And Want To Buy Again', Titijaya is now highly regarded as an innovative, resourceful and reputable property developer, responsible for many successful developments in the most sought-after growth areas. His children, Poh Yit and Charmaine, have already expressed their interest in the property development industry by being involved in the family business. Poh Yit is currently the Deputy Group Managing Director of Titijaya, while Charmaine is an Executive Director of the Company. Apart from

M. Jaafar et al., *Business Sustainability Model for Malaysian Housing Developers*, DOI 10.1007/978-981-10-5266-8_17

being a successful businessman, Tan Sri Dato' Lim Soon Peng is also a humanitarian. He is the Executive Advisor to the Federation of Malaysian Lim Associations and was solely responsible to secure a grant from the federal government to help Malaysian Chinese women with low or no income in the form of low interest-bearing loans.

17.1 Firstly, We Would Like to Know Your Personal Profile. Could You Please Tell Us Your Family Background and How Did You Manage to Enter This Industry?

I was born in Kapar, Selangor, and my hometown is at Pandamaran Jaya. We used to stay in a squatter community there. I am the eldest child of 12 children. My parents and forefathers came from China. They were so poor there and decided to move to Malaya in order to change their life. That is why I always say that I came from humble beginnings, having to leave primary school in 1967 to help full-time in the family business which included pig rearing and petty trading. I only had 5 years of schooling and began working at the age of 12 at the Klang Market as a labourer – loading and unloading items. When I turned 17, I decided to take up construction skills as my job at the market was only earning me about RM50.00 a month. In those days, there were no concrete machines, and I had to learn to mix and produce concrete manually. Together with two friends, we set up Chan Yik Enterprise, a subcontracting company, in the early 1970s. This was my stepping stone into the construction business and subsequently the property sector. I eventually established my own company, Titijaya (M) Sdn Bhd, in 1983 to undertake property development, mainly in the Klang area. Through this company, I successfully developed my first project comprising holiday apartments in Fraser's Hill through a joint venture with the then Malaysian General Investment Corporation Berhad. I made my first million Ringgit at the young age of 28, from the building of the holiday homes.

17.2 How Did You Manage to Secure the Project at Fraser's Hill, Pahang?

The joint venture project between the Pahang State Government and Malaysia General Investment Corporation Berhad was awarded to me when other contractors were not keen to tender due to fear of higher cost of transportation and building materials as a result of the road conditions at Fraser's Hill. Nobody wanted the job as it was not easy transporting materials up and down the hill. The construction at Fraser's Hill posed a number of challenging factors such as location, earth suitability, logistics and so forth. Nonetheless, I was determined to successfully deliver my first development project of holiday apartments, and fortunately, the project took me

3 years to complete. With the money I made, I bought land for small project development like Taman Bukit Cheras. I also invested in oil palm estates in Meru, Klang.

17.3 When Did You Start Your First Project?

We started out with smaller projects in Bukit Raja, Klang. It was during the economic crisis in 1998 when land price was low. With the support from banks, I decided to delve further into this industry. My knowledge and experience in the property market was basically my own efforts and achieved success at a young age, working step by step. My greatest financial gains were made during the 1998 financial crisis. Our earlier projects involved building of double-storey homes and three-storey shop offices in Cheras, while in Klang, we built low-cost apartments and double-storey homes. From then, everything was going great for me as Titijaya had since grown by leaps and bounds, developing some of the most prominent real estate projects in the region. It was in 2004 that we rose in prominence when we embarked into our first high-profile development project where we launched the E-Tiara Serviced Apartments and the Klang Sentral Commercial Centre in Subang Jaya and Klang, respectively.

17.4 Why Did You Dabble in This Housing Development Industry?

I take cognisance on the importance of owing a house by everyone. If we have the capabilities and interest, we should take on and uphold the responsibility to build houses for the people. We are not here solely for profits but we want to help low-income people in particular to own houses. That is why we form organisations and societies to assist them. I used to be in their shoes, and this indirectly pushes me to help the needy.

17.5 What Are the Differences Experienced by This Company When It Went Public?

The soul is still the same, as we are still into property development. But we did expect some changes after Titijaya Land Berhad went public in terms of some newer ways of doing things. After our listing, the investing public will be expecting us to grow, and so we cannot limit our growth towards a certain range of products. We have to keep moving forward. This is an opportunity for the company to embark on other segments of the market.

17.6 In Order to Succeed, What Traits that You Think Should Be Imbued by Housing Entrepreneurs? Do You Mind Sharing with Us Other Success Factors that You Deem Important to Excel in This Industry?

I would like to confess that there are no substitutes for hard work together with careful detailed planning. I emphasise a lot on the housing concepts to add value for our property buyers, and I encourage my staff to embrace building trust in doing business. On a related note, we must be aware of the current changes in our surroundings. If we know that there will be new plans for infrastructures or a transportation hub near to our land, we must quickly grab the opportunity. Other factors such as current demand, population and demographic profiles are vital to be studied. To a large extent, we have built bus stations at our projects in Klang and Seri Kembangan to attract buyers and make the locations livelier. This approach will assist youngsters to make the decision of owning their first house rather than buying cars as conducive public transport is provided for them. Next, we also notice that buyers nowadays are getting wiser as they prefer to buy houses near to public amenities such as school, hospital and LRT station. This can be seen in bigger cities like Kuala Lumpur and Penang where a number of condominium buildings have been built near to a hospital in order to cater to the demands of the elderly. This new observation was obtained from our recent research, and we have decided to start investing on new land banks that are located near to public amenities.

Apart from that, I believe that entrepreneurs should always meet people to widen their network. We may need to engage consultants to inject new ideas and advice for our projects. Indeed, we must look on the current market trend and demand. Housing developers should have updated industry knowledge especially on external factors and technical know-how. In view that land price in the Klang Valley is escalating, we may need to emulate other developers to build more high-rise properties. The scenario is different in Banting and Klang, where the land price is relatively lower and the demand for landed properties is still overwhelming. As for my own strategy, we will only launch one project a year. In fact, we build our houses phase by phase. The idea is to minimise the risks that may occur along the way of construction.

17.7 Take the Example of Subang Park Homes, How did You Acquire the Land?

We purchased the land by tender. I knew when we acquired this land almost 10 years ago that it would be a secure and exclusive development. It is like an island, near the town centre but exclusive as we have our own private access road leading to the development. At the moment, Titijaya's 600-acre land bank includes tracts in Subang Jaya, Meru, Kapar, KLCC area and Penang.

17.8 Marketing Is Very Important in Business. Do You Have Any Special Methods to Market Your Properties?

Before we implement the marketing process, we will first determine the target group. The proposed design should be parallel with the acquired demographic profile of the area. Location of the land must be strategic. We need to seek for the buyers that intend to purchase our products. Before we launch the projects, our staff will carry out studies on the marketability of our properties and strategise accordingly which will encompass preferred design, target group, etc. We are concentrating on high-end residential products and medium-cost houses that require us to prioritise on quality. Our focus group is more on youngsters and new families who wish to have their own houses. Perhaps, in the future, the approach will be different as we may shift to focus on the elderly group. Again, it depends on the project location. What sets Titijaya apart from the other property developers in Malaysia is our focus, which is more on small-scale while others concentrate on bigger-scale developments. We focus on areas that are more easily accessed by public transport, for example, areas near to public transportation facilities, like Subang Jaya, which is close to a train and an LRT station. This is a convenience for the young people. Our strengths lie in doing businesses that cater to younger people.

17.9 Could You Please Elaborate on the Managerial Procedure of Your Company?

We monitor our staff performance through their heads of department. The assigned managers will report to me every week to keep me updated with the current issues. As for our project matters, I believe that we need to have cash reserves on standby either through external or internal resources. If such reserves are not available in hand, our projects may face challenges during the development period. This strategy is quite similar to insurance coverage. If we are faced with any problem, the insurance will automatically help solve the predicaments. On a separate note, we have also adopted the use of ICT in our organisation, particularly by our marketing division. I encourage the use of social media, for example, Facebook and other Internet sources, to boost our sales.

17.10 At the Moment, Do You Have Any Specific Strategies for Your Business?

As I have mentioned before, our vision is to expand the business all over Malaysia especially Penang, Johor, Sabah, etc. However, in order to stay competitive, we have to firstly strengthen our brand name locally. We have applied some innovations to our projects in tandem with global promotion towards green products. Green Building Index (GBI) has been taken as our basis, and we have implemented it in our project at Seri Kembangan. For our next project, we plan to utilise solar concept to reduce the use of electricity. This technology is in high demand overseas and is supported by our government. For me, we need to follow and be always updated with this new technology. Notwithstanding the above, we plan to concentrate our projects to big cities only.

For instance, 3Elements@Puchong South is our first Green Building Index-certified project, and we plan to continue developing more green buildings. We want to educate and create awareness among our purchasers on the benefits of green buildings on a long-term basis, especially cost savings on electricity, as well as a healthier and more comfortable living environment through better ventilation. We will also look into our products' function ability especially on the layout and design, better ventilation and less usage of air conditioners. The properties developed by the Titijaya Group are aesthetically conceptualised with elegant landscaping for a modern and contemporary lifestyle. These details are manifested in all our ongoing and upcoming projects.

17.11 Please Brief Us on Titijaya's New Projects Which Are Scheduled to Take Off Soon

Indeed, any future forays into the high-end market will be dependent on location and purchasing power of the target market. We plan to develop a mixed integrated project with a gross development value (GDV) of RM2.5 billion on a piece of land located in Jalan Eaton, Kuala Lumpur. We have also ventured into Penang after acquiring an 8.1 ha parcel near the Penang second link for RM126 million. Last April, we also signed a joint venture agreement with Bina Puri Construction Sdn Bhd to develop the RM1.3 billion mixed development in Brickfields. The proposed project will consist of serviced apartments, to be targeted at the mid- to high-end market segment, as well as a hotel as one of our diversification plan. Besides that, we have undertaken a few ventures in Sabah which will generate substantial GDV for the company.

17.12 Does Titijaya Involve in Any Other Industry, Apart from Housing and Commercial Developments?

Titijaya is also developing an industrial project called Zone Innovation Park on a 100-acre freehold land in Kapar, Klang. We sell subdivided lots of about half an acre each or according to the needs of our clients. The industrial park, which was opened for sale in mid-2011, has a total of 101 lots priced from RM2.13 million. Another similar development is the RM132 million project called Seri Alam

Industrial Park, located in Klang. Apart from that, we also have an aluminium factory under a related company, Kamco Aluminium Sdn Bhd.

17.13 What Do You Think of the External Forces of This Industry? Do You Think Government Intervention, Macroeconomic Factors and Competition Affect Your Business Performance? Are There Any Related Challenges that You Want to Highlight?

As you may know, the government has its policies. Our development procedures require the government's assistance by expediting the project's approval. Global economic conditions do not affect the housing industry very much. The Klang Valley will always be populated as people will be coming from other states because of job and business opportunities. They will come here to work and live. Local property prices attract a lot of foreign investors. Our prices are still the lowest when compared to those of neighbouring countries. Noticeably, properties in Malaysia have been well received by external investors since the last 10 years, and this has greatly contributed to the country's economy. Challenges are plenty and I would expect that Klang Valley will suffer from water problems in the near future since the development growth has been tremendous.

17.14 We Notice that the Management of Your Company Has Been Spearheaded by Your Children. Can You Please Clarify a Bit on This Approach?

Actually, it is their interest to want to learn the intricacies of this business. They have adequate competencies and education to improvise the policies and leadership of the organisation. Their advanced knowledge and exposure to the Internet have allowed them to gain new knowledge and ideas about this property business. Indeed, I see it as a way of expanding my company. If I were to continue doing this business on my own, I would not be able to expand it to this size. My children have an interest in the business, and so we will expand the business and bring it to greater heights. We have to let go little by little and let the children take the lead so that the business can grow further and bigger. We have to move a little to the sidelines. I let them have a deciding hand in many matters, and I do not want to continue overseeing their work. If you do not let go of the leash and let them do it, they would not be able to learn anything. We will see the outcome when decisions have been made, and if mistakes are made, I will take a look at them. Through this, I will be able to assess their abilities. When my children are used to making such decisions in the business, then they will take the initiative and be independent. However, if you grasp them tightly and do not give them leeway, they would not be able to grow up. Around 50–60% of my business has already been relinquished to their control. I have told them many times to have courage to decide before I take a look on it. Although I may be open to my children following their own dreams, I still believe that this business will always be a rock, a firm foundation for them. Nonetheless, I am still the final decision-maker of the company.

17.15 When Were You Conferred the Titles Tan Sri and Dato'? Do You Mind Sharing with Us the Other Awards That Were Received by Titijaya and You Personally?

I was conferred the title Dato' in 1994 while the title Tan Sri was conferred in 2010. That same year, I was awarded the title the Best Philanthropic Private Developer at the inaugural NST Property – SC Cheah Choice Awards. I won the best philanthropic award because the amenities that built by us were for public use without charge. These include schools and bus terminals. This is our way of giving back to the society. More recently, I was awarded the BrandLaureate Brand Icon Leadership Award in 2014 at the BrandLaureate Awards Ceremony. As for our projects' accolades, Titijaya was the winner of the prestigious Asia Pacific Commercial Property Awards – 5-Star Best Mixed Use Development Malaysia 2010. Our project named First Subang was accorded the Highly Commended Office Development Malaysia Award by the International Property Awards in association with Bloomberg Television in 2011. Another project, Subang Parkhomes, also obtained the Highly Commended Apartment Malaysia Award at a similar occasion. Prior to that, One SOHO, another outstanding project in Subang Jaya which was launched in 2008, won the Silver Award for the Residential Interior Category of the MSID Interior Design Awards in 2008. It was also awarded the Best Mixed Use Development Malaysia at the Bloomberg Television Asia Pacific International Property Award 2010 held in Hong Kong in 2010. Basically, we are proud with our projects for these continuous recognitions.

17.16 Could You Please Provide Some Words of Wisdom That Would Be Useful for Young Entrepreneurs Who Wish to Delve into the Housing Development Industry?

We always look at other nations that have already advanced, such as Hong Kong and Singapore, while China has substantially improved in its economy. I posit that we should embrace their strides and ideas, so that we too can transform according to our own scenario. When I started doing business, I had problems communicating with others as I cannot even speak and understand Malay or English. But gradually, through mingling with many people, I have managed to learn the languages and take up projects with people of other races. We need to be humble and honest with them as they tend to be understanding towards our weaknesses. I personally take this barrier as a challenge for me to strive for success.

Chapter 18
EUPE Corporation Berhad

Keywords EUPE Corporation Berhad • Beh Huck Lee • People oriented • Market study • LEED certification • Stakeholders' trust

In nearly 30 years, EUPE Corporation Berhad has built more than 20,000 homes, comprising 5 million square feet of residential and commercial property space in the northern Malaysian state of Kedah. To date, it has undertaken seven major developments in Kedah, with a combined GDV of over RM1 billion. EUPE Corporation Berhad comes from the name of a rubber estate, i.e. UP Estate (United Plantations), where it had built its first development in Sungai Petani – the state's largest population centre. For almost three decades, EUPE Corporation Berhad has been a trusted name behind some of the notable projects in Kedah. It has now been expanding its footprint into Malaysia's key property markets with major developments in Kuala Lumpur and Penang Mainland.

The company was founded in 1986 by Datuk Paduka Beh Heng Seong and Dato' Wira Tajudin Hashim during the recession of the mid-1980s. It was established in response to the state government's call for the private sector to meet the housing needs of the low-income group. This led to EUPE's first major project – Taman Ria – which was then the largest residential development in Malaysia with over 3000 units. Today, EUPE Corporation Berhad has divisions in property development, civil engineering and construction, property management, property investment as well as hotel and golf course operations (through its Cinta Sayang Resort and Hotel in Sungai Petani). EUPE's Managing Director, Dato' Beh Huck Lee, holds a Bachelor of Commerce and a Bachelor of Engineering (First Class Honours) from the University of Western Australia. Prior to joining the group, he was attached to Hewlett-Packard. Taking over the group in 1995, he consolidated all 13 of its companies into its present form as EUPE Corporation Berhad. When the company was listed on the main board of the KLSE in 1997 with a paid-up capital of RM128 million, Dato' Beh Huck Lee became the youngest Managing Director of a listed firm in the country, at age 27.

A father of two, he has been involved in the business for 20 years now. Initially, he focused on mass housing developments in Kedah, but the company is now focussing on low-density, luxury projects that are environmentally sustainable. This approach can be seen in Sky Residences, the group's high-rise condominium build-

M. Jaafar et al., *Business Sustainability Model for Malaysian Housing Developers*, DOI 10.1007/978-981-10-5266-8_18

ing in Cinta Sayang, which was the first building in Malaysia to be awarded the Leadership in Energy and Environmental Design (LEED) certification by the US Green Building Council. Other flagship projects include Astana Park Homes, The Somerset, Bandar Seri Astana and Alma Hillpark in Penang. Its first two projects in Kuala Lumpur are called Novum, located in South Bangsar, and The Weave in Cheras. He received his Datukship in 2011 from the Sultan of Kedah.

18.1 How Did This Company Get Started?

The company was set up by my father, Datuk Paduka Beh Heng Seong, along with his friend, Datuk Wira Tajudin Hashim. They have been friends for so many years since Datuk Wira was a Kedah civil servant. Prior to setting up EUPE, they had undertaken a few small projects together. In 1986, the national economy was recovering from very bad economic crisis. Malaysia's then Prime Minister, Tun Dr. Mahathir Mohamad, wanted to start a special housing project in the region and was supported by the then Kedah's Menteri Besar, Tan Sri Osman Aroff. The intention of the government was to build low-cost houses for the people with at least two separated bathrooms (minimum 750 square feet). This led to the Taman Ria project, comprising 3300 houses with the majority of them being low-cost houses.

18.2 May We Know a Little Bit about Your Father's Background?

He is from Sungai Petani. Before starting EUPE, he was helping my grandfather with their family food manufacturing business. During those years, he picked up knowledge about the construction and development through community work he was involved in. For example, he was involved in the building of a temple, and through this, he learnt a lot by being a site supervisor and carpenter as well.

18.3 How About Your Own Background?

I finished my commerce and engineering degree at the University of Western Australia. It was a combined degree, more akin to double degree, over 5 years, from 1989 to 1993. After my graduation, I worked for Hewlett-Packard in Penang for about a year. As the eldest child, I took full responsibility to continue on my father's business after my father entered politics in 1995. Thus, he needs to relinquish all his shares. At that time, we do not have many choices because if I did not take over, we need to seek other new partners to manage this company and they might take over all the shares altogether. That was the beginning of my life here.

On the other hand, for your additional information, Datuk Wira's family is still the partner of this company. His son, Faisal, is with us and actively involves in the management of EUPE.

18.4 Please Elaborate on Your Flagship Project, Taman Ria

Through phase-by-phase strategy, the whole project was completed within two and half years only, which include delivery period. Taman Ria is one of the biggest residential projects in northern part of Malaysia, about 340 acres in size. When we started Taman Ria, we were among the first few developers in the country to build a retention pond because the project is quite sizeable, in order to avoid natural flooding. We were proud to see Tun Dr. Mahathir Mohamad took a helicopter from Alor Setar, to check the quality of the development. After that point of period, Taman Ria was used as a study case by a lot of state economic development corporations (SEDCs) around Malaysia. From a small beginning with our first project in Taman Ria, the company has now built a portfolio that includes the biggest mixed development township in the state and commands a majority of the market share in the Sungai Petani area.

18.5 Could You Please Explain to Us How Did You Build Up This Company After You Took Over the Office?

When I first took over the company, I found that there are a lot of areas which could be improved, and I also saw that the industry environment was changing. In the past, we were running the firm as a family-owned business which means the owner must be hands-on in every single area. To move forward as an organisation, we need to be corporatised and structured, and employees should be able to determine their own path of growth. At the same time, they must share the same vision with the organisation. Traditional businesses do not allow for that kind of clarity, direction and vision.

So firstly, I got in touch with merchant bankers to get the firm listed. This would allow us to tap the financial market more efficiently and build the company immediately. We were listed in 1997, fortunately, before the Asian financial crisis.

18.6 What Are the Characteristics that You Think Are Important for You to Lead This Company?

When I took over the administration, I used to spend a lot of time focusing on the details of every operation. But these days, for the organisations to grow, we must be able to get the right person and set in place the right frameworks and principles for the organisations to operate. I disagree with traditional assumption whereby the employer is the sole mastermind behind every single success. If you want people to work successfully with you, you must also pursue the interests and principles which are important to them and listen to their ideas too. That is very important. The other key goal is for me to set key priorities in order to guide the staff to hit a common target. Suffice to say, business owners must prioritise business strategy and the growth of human capital since business success depends much on the capability of human resources.

18.7 How Does EUPE Come Up with Diversification Strategy in Business?

EUPE aims to become the most trusted brand in every sector in which it operates. Currently, we have five key divisions, and they are interlinked or related to each other. Property development is the mainstay and then construction, property investment and management, golf and hospitality division as well as the plantation division. Diversification has evolved as a strategy to meet different challenges in the company's development as well as maximise new opportunities. For example, when we built Taman Ria, it was our responsibility to provide the residents with basic facilities, such as private-owned market to cater the community needs. From there, our property management and investment division was set up. Our golf and hospitality division started Cinta Sayang Resort back in 1988 to take advantage of the growing market for leisure and hospitality services in northern Malaysia.

18.8 How Do You Acquire the Land Bank Specifically in Other States?

Besides directly purchasing from landowners, we also buy land from state corporations. In Kedah, we currently have almost 700 acres, and we are actively searching for new land development opportunities in the Klang Valley.

18.9 How About Your Overseas' Projects?

On the global front, the company has a presence in Australia. Last September, we had announced that our 70% owned Australian subsidiary, Australasia Development Pty Ltd., had set up a 50% owned company, The Surrey Sunbury Pty Ltd., as part of our expansion plans. At this moment, my purpose in investing there is for strategic purposes rather than profitability. We are there because we want to diversify our risks and grow our organisation in terms of business skills and profile.

18.10 Looking at the Recent Market, How Do You Cope with the Changing Market Preferences?

For different projects, we have different market profiles. Take the example of Bandar Seri Astana; it is a new development area. Thus, we have to provide proper infrastructures during the starting point in order to get proper connection to nearby estates. That was the first challenge – to have proper linkage and access. But, through the assistance of government agencies, we were still able to develop the land. On the other hand, for our resort homes project, we noted that the buyers want to get closer with their workplaces since we have many factories nearby. So, we design our houses in accordance with their profiles, age and preferences. You may see that in Sungai Petani itself, different patterns can be discerned. Different markets pose diverse demands.

For Astana project, we have few parcels, i.e. Astana A, B, C, D and E. Parcels A, B and E are more on mass housing areas, while parcels C and D are slightly different as they supply for uptrend market. Again, we need to look at the needs and network distribution. For example, parcel C is very near to OSK Commercial Centre; thus, we need to do a lot of upgrading works especially on the road. Also, some of the areas are slightly more hilly, and we do not want to cut so much. If we want to do all mass housing concepts, we need to flatten all areas, and it can be very inefficient.

18.11 What About the Implementation of Goods and Services Tax (GST)?

I think GST is unavoidable because many countries have already implemented it. It does introduce some uncertainties during the initial stage as consumers are trying to adjust for it. But since other countries have successfully applied the scheme, I believe people will get used to it, and the system will readjust itself back to the equilibrium. It has had some short-term impacts on housing markets, but the effects will soon be recovered.

18.12 We Perceive that a Lot of Housing Projects Are Mushrooming in Sungai Petani, How Did You Manage to Steadily Sell Your Products?

The demand pattern here is quite resilient since the houses are still comparatively quite cheap. But there is a strong and growing market for higher-value products. Right now, we are selling our terrace houses at RM300,000 per unit, but the sub-sale can be almost double. For semi-Ds, we used to sell below RM500,000. Today, the sub-sale price can already go from RM900,000 to RM1 million.

18.13 How About Your Sky Residences Project? Who Are the Target Buyers for This Project?

Properties here in Sungai Petani are relatively cheap, and a lot of families here have big houses. Sky Residences are targeting elderly groups who might want to move from their big houses to the serviced apartments, and we believe that this project has huge potential to attract local and foreign retirees. Sky Residences is located near the hotel operation, and the residents also have access to better catering and services. With our plan to expand our business in Kuala Lumpur, we also wanted to make this project an important learning platform.

When we first launched it, we believed that Korean tourists might be interested to buy our serviced apartments since they tend to spend 3 or 4 months here to play golf at our Cinta Sayang Resort. Surprisingly, our forecast was accurate when 35–40% of the occupiers are Korean. In addition to that, we have a group of semi-retirees comprising professionals, like doctors and lawyers, and they have form good community among themselves.

18.14 Is It Your Own Initiative to Build a Water-Themed Park?

We signed with PKNK (Kedah State Development Corporation) to develop a piece of land near Cinta Sayang Resort for housing project purpose. But due to economic turmoil and because the golf course business was deemed irrelevant anymore, we had turned it to other type of property development. We were approaching Korean tourists at that time; that is why the idea came up. They like water and same goes to Malaysians. A water-themed park is more cost-effective than other types of leisure-themed parks. On top of that, water-themed park produces less problems in terms of management and maintenance.

18.15 How Much Is the Profit of This Company?

For the last few years, our gross profit for property development was between RM25 to RM40 million. Our projects in Kuala Lumpur will enhance our portfolio and returns quite substantially. Comparatively, if we go to Kuala Lumpur with small and niche project, it will be equal to the 100-acre project in Sungai Petani.

18.16 How Do You See the Importance of Different Departments in Your Company? Which Departments Would You Like to Emphasise for Contributing to the Firm Success?

I would say that each department plays a major different role in every stage of value creation. Before the project launch, marketing is the most important division. Once the project is launched, the sales department takes the lead. Subsequently, finance and construction sides go hand in hand, while cost control is very pertinent towards the end of the delivery in order to prepare feasible information for our next project. Human resource department is essential at all times. In general, they are equally important but during different time.

18.17 Can You Share Your Marketing Strategy?

One thing I would say is that we are very people-focused. We try to introduce innovation before others. If we look on Sky Residences, we are the first developer in this country to get LEED certification. I would not sell the property that I myself do not want to buy. So, to answer your question, I would say that effective marketing is all about knowing what your customers want and giving what they want. Consider cost comparativeness wisely, provide best services and facilities frequently, and try to think out of the box at all times.

18.18 What Are the Overriding Strategies to Compete with Other Rivals?

We are quite lucky that our alliances put huge trust on us. Our sale speed is usually quite fast compared to our competitors. We have been working with our contractors for a while, and we know what are the critical areas that we need to control. Our suppliers are comfortable with us as they offer us relatively competitive rates. Even during the material shortage, we were given priorities in delivery. Government agencies also are very supportive as we are able to deliver what we have promised in serving the community.

18.19 Do You Apply Any Innovation in Your Products? What Type of Innovations, and How Is It Applied?

At that period of time when we were looking for green certification, we do not have GBI or Green Mark or anything for a yardstick. We shortlisted two standards – LEED from the USA versus BREEAM from the UK. We opted for LEED because it emphasises user comfort as well as being green. For this innovation, we had engaged Surbana International Consultants from Singapore to advise us.

18.20 Since Your Company's Vision Is to Be a Trusted Brand in Housing Development, Could You Please Elaborate More on the Importance of Reputation Towards Business Success? Are There Any Other Qualities That You Think Have Made the Buyers to Purchase Your House?

Building trust links our five key stakeholders – consumers; partners; alliances such as contractors, lawyers and bankers; potential employees; and shareholders. We always try to build trust with all these people. In whatever we do, we will take long-term perspectives to build trust.

18.21 How About the Use of ICT? Is It Useful in Your Marketing Strategies?

ICT is getting more prevalent and significant in marketing. Since the information is easily disseminated nationwide, we manage to accumulate repeating buyers. They do not come back to merely buy our properties but also bring new acquaintances to buy our products. I believe ICT also plays a role in this sense.

18.22 Currently, Many Housing Developers Lament that the Housing Loans for Homebuyers Are Very Hard to Be Approved. Do You Agree with That Situation?

Banks are becoming more stringent. This scenario is in tandem with the consolidation of what has been a long period of growth in property prices. Previously, the rejection rate is less than 10%; now it rises to 30%. Of ten potential buyers, we shall be prepared that only seven of them are eligible to buy our houses. But I think, sooner or later, the policy will be changed a bit because it affects the industry. Malaysian housing market needs to be stabilised first before it can seek for new directions of economic development. Hopefully, it would not go too fast. We have to be very careful on what we offer, and we need to push our comfort boundaries against all odds. It is true that networks, brand and trust are important, but please bear in mind not to take things easily and for granted because it may jeopardise your business.

18.23 From 2008 to 2013, Kedah Was Ruled by the Opposition Party, and Now It Returns Back to the Federal Government, Are There Any Differences Working with a Different State Administration?

Not really, because land law that covers our operations is already established and enforced. Also, the administrative work is fixed with no intervention from the politics. Obviously, there are differences in overall goals of both governments. The current government emphasises on economic speed-ups, while the previous one focuses on catering the needs of lower-income groups. Both objectives have pros and cons, but in the end, political and policy environment is only one of the myriad factors that affects property development industry.

18.24 Please Share with Us the Overall Business Approach that You Have Utilised and How It Relates with Your Organisational Values

Our overall business strategy is different from other players in the industry. We do not just buy a piece of land, have it converted and pass the risk of construction entirely to the third party that will eventually trigger many uncertainties in terms of speed, quality, cost and learning. That is why we started our construction arm since 1987 to ensure that we can manage the uncertainties effectively for our customers.

In general, we pride ourselves through the emphasis of our main corporate value that entails trustworthiness. We prefer to take the long-term view rather than chasing the short-term profit. To gain trust from the buyers, we make sure that we are constantly designing and developing properties that enhance value for both buyers and the community.

18.25 Do You Have Any Lessons that You Are Keen to Share with the Nascent Developers that Are Interested to Thrive in This Housing Industry?

My best piece of advice is not to give up easily. The industry can often be very challenging, particularly if the market is slowing down. With many issues to manage, especially regarding ambiguity and vagueness, you have to hold firmly on your principles and values. I have my own team and colleagues who are very helpful and supportive, supplying me with good advice. The other important lesson is not to treat your business rivals as enemies. Indeed, there is always something to learn from your fellow competitors.

Chapter 19
Koperasi Tunas Muda Sungai Ara Berhad

Keywords Koperasi Tunas Muda Sungai Ara Berhad • Mansor Musa • Cooperative society • Collective decision • Sterling reputation • JV strategy

The history and challenges to build up Koperasi Tunas Muda Sungai Ara Berhad (KTMSAB) were not without its fair share of obstacles and issues besides the enormous contributions and sacrifices from individuals and support from other organisations. KTMSAB was founded in the year 1963 by several local leaders with the support from UMNO Sungai Ara, guided by Mr. Jaafar Abd Majid, a well-known Malay cooperative officer in Penang. But, it was only in 31 December 1966 that KTMSAB managed to be registered under the name *Syarikat Bekerjasama-Sama Serbaguna Tunas Muda Sungai Ara* which was later amended to Koperasi Tunas Muda Sungai Ara Berhad on 30 September 1980. As of 1 January 2012, a total of 1296 members was recorded with the racial composition of 927 Malays (71.2%), 341 Chinese (26.3%) and 32 Indians (2.5%), respectively. With a vision towards becoming the country's leading property developer by 2020, this organisation owns three subsidiaries, namely, Sri Tunas Sdn Bhd (in construction), Tunas Harta Sdn Bhd (in real estate) and Sri Tunas Edar Sdn Bhd (in trading), to facilitate its main business. For the record, it is the best cooperative enterprise in Malaysia under the real estate category and was placed fifth among top Malaysian cooperatives based on the index issued by the Malaysia Cooperative Societies Commission in 2013. Among the notable residential projects of this housing cooperative are Taman Sri Tunas, Taman Tunas Muda and Taman Tunas Damai. Currently, the Chairman of KTMSAB is Datuk Ir Mansor Musa who has willingly shared some of his opinions and experiences, documented as follows.

M. Jaafar et al., *Business Sustainability Model for Malaysian Housing Developers*, DOI 10.1007/978-981-10-5266-8_19

19.1 May You Please Brief Us About Your Personal Background? How Do You Manage to Lead This Housing Cooperative?

I graduated from Madison Wisconsin University, USA. At first, I worked in Penang for 2 years and another 2 years in Langkawi, Kedah. Then, I decided to return to Penang. I was involved in this cooperative society because this is my hometown. In fact, my wife is also originally from here, while my parents-in-law are also members of this cooperative. Experience-wise, I am a civil engineer by profession. I used to work with a contracting firm, consulting company and the government prior to my tenure in this cooperative. Prior to becoming the Director of the cooperative, I had a chance to become the people's elected representative in this area and have my own contracting business. Obviously, my entire background is in the construction sector. It thus moves in tandem with this housing cooperative that deals with construction facets. My background helps me a lot to monitor and manage this organisation. Firstly, as the Non-Executive Chairman, I need to chair the General Meeting where all the policies require endorsement by the Board. We have appointed a General Manager (GM) to head the company with more than 60 staff to run this organisation. The GM will only implement the decisions made by the Board. For instance, if the Board agrees that certain areas should only be developed for the sake of our members while other parcels can be sold for generating profit, the GM needs to abide by this order. After all, each decision will only be executed once the Board has had a discussion with the management. That is the main task of the Chairman.

My father became a member of this cooperative in 1966, and until 1999, as age caught up with him, our family had a small discussion and decided to transfer the membership to my eldest brother. Well, this situation happened prior to that, somewhere around 1995. But from 1999 to 2000, I was active with the political party, UMNO, and other associations as well, so, we decided to transfer the membership to me instead. I thus became a member of this cooperative since 1999. After 2 years, I chose to participate in the election of the Board of Directors and was elected as one of the members of the Board from the year 2000 to 2008. In 2008, I was promoted as the Chairman of this organisation. For your information, the Chairman is entrusted to oversee the day-to-day operations and is responsible for the management and marketing aspects of the corporation. To that extent, I bring forth my expertise of working in the engineering and construction industry to facilitate the operational part of the business.

19.2 What Is the Process of Electing the Board Members?

The Board of Directors are elected during the General Meeting. Every year, four members will be selected to join the Board. For your information, the Board consists of 12 members, and each of them will undergo 3 years of service. After the tenure expires, they need to undergo the similar process of re-election. Immediately after the annual general meeting, the Board of Directors will elect one chairman, one secretary and one treasurer to handle the administrative and managerial procedures.

19.3 Do You Have Any Specific Mentors that Inspire You to Excel in This Field?

Frankly, no particular persons since I socialise with everybody, i.e. politicians, businessmen and others. It is good to have a working mentor, but for me, it is not necessary because my father was just an ordinary labourer in the Universiti Sains Malaysia (USM), if we are taking the example of my family. The most important is how we deal and mingle around with others. More often than not, my background in the construction has helped me a lot.

19.4 What Is the Vision of This Organisation?

We wish to become as big as S P Setia Berhad, Mah Sing Group Berhad and Sunway Berhad, the real estate juggernauts in Malaysia. We do not want to only clinch the 'number one' position in the cooperative segment. We hope to become among the top developers in Malaysia. It is just the problem with the decision-making process that hampers our progress since our administration is subject to the Malaysia Cooperative Societies Commission (SKM). Also for the construction industry, there is a lot of 'under-the-table payments' in order to get the approvals especially when we deal with projects worth hundreds of million. This is the reality in Malaysia, although we might have achieved the standards on paper.

19.5 Please Tell Us on How This Cooperative Developed Its Business in This Real Estate Industry

It started in 1966 when the members bought the land bank of 1007 acres worth RM2 million. At that time, this land had no real value at all. In 1970, the state government purchased 260 acres from us worth RM3 million. That enabled us to settle our debts, quite apart from getting a million profit margin; it also left us with 747 acres of land bank and more than RM800,000 cash in hand. That was how the cooperative was initiated. During the 1980s, factories were mushrooming in Bayan Baru (near to this area) that drove the housing demands. It was the beginning mark of our progress in the property sector, erecting and selling houses with significant profits due to land availability that we had acquired at practically no cost. Now, I would say that the property market here is booming extensively. In 2008, the value of land in the area was recorded at RM30.00 per feet, but now it has already surpassed RM100.00 per feet. With this huge windfall, we became extremely profitable and managed to build houses for our members for free.

19.6 How About Your Cooperation with Sunway Berhad to Develop Housing Estates and Collaboration with Pantai Medical Group to Build a Private Medical Centre at Your Area?

In the 1990s, Pantai Group approached us for a joint venture to build the private hospital. We provided them with a parcel of land, and they gave us 30% of shares in the form of equity. But after a few years, acknowledging that we have no expertise in this field, it prompted us to sell our shares. Separately, the alliance with Sunway Berhad was also embarked in the 1990s. This smart partnership with an experienced private property developer in certain strategic areas was deemed imperative to enhance our job efficiency. You might be aware that cooperatives usually lack decision-making skills, resulting in ponderous outcomes and slow judgement. Another thing is while we are having a number of projects in hand, we decided to apply the joint venture (JV) strategy to overcome the problems of limited human and financial capital with precarious exposure that is may be high and beyond our grasp. If we have development projects that value almost RM1 billion, the exposure will be extremely high. The JV strategy allowed us to avoid uncertain predicaments during market collapse. Currently, some of the projects will be carried by our own team, while others might be collaborated with other parties. We are now in JVs with local developers such as Ideal Properties Sdn Bhd, SEAL Incorporated Berhad and Syarikat Nyaman Sdn Bhd especially for high-end housing products. If we wish to set up a JV with other developers, we will ensure that we are dealing with direct personnel of such firms, not via brokers. Our procedure for JV is to ask for RM10 million upfront as deposit for guarantee. If they do not trust us, they would not participate. The deposit acts as our safety net, and when we execute this method, it is anticipated that our projects will not confront funding problems and be completed on time.

19.7 How About Other Land Banks in Penang?

For all our land banks in Penang, we have already submitted several proposals for the layout process, to be considered by the local authority for future developments. At the moment, we have about 60 acres that are undeveloped, but, still, we will do so in the near future as it is under planning.

19.8 Can We Know About Your Market Focus? Who Are Your Target Markets?

In Penang, we keep an open-market concept as there is no such thing as Malay Reserve Land. For us, we build all types of houses that comprise expensive high-end products, as well as affordable and low-cost houses. The percentages are rather equal for all types. Basically, we are subsidising these low-cost and affordable houses that cost us about RM42,000 each. In every housing scheme, there must be a 30% allocation for the provision of low-cost houses.

19.9 While You Dealt a Lot with the Vagaries of the Board of Directors, What are the Main Problems that You Need to Overcome?

I acknowledge that it is difficult to deal with human beings. But as a politician, I found that it assists me a lot since I have experience in human relationship. As for me, the connection with the Board of Directors must be tight and warmth. The mutual trust and tie should not be confined in the meeting room only; it needs to transcend beyond normal bond. You must treat them as your good friends. With this approach, decisions and agreements made will be easily achieved.

In this type of entity, it is not easy to get everybody or the majority to agree with certain proposals as it entails various thoughts and ideas. The decision will not be executed when half of the members expressed disagreement. Normally, if five of the members (out of 12) disagree, we will analyse and review the proposal again for better and more positive outcome in the next meeting.

19.10 How About Staff Management? Is the Construction Site Handled by Different Workers?

While there are certain jobs that we perceived to be lacking in, we will certainly outsource the skills. In terms of subsidiary, I think it is not advisable for us to expand because the staff tend to be complacent with their environment. They take their jobs for granted, resulting in low productivity. Some lack competitive spirit since they know that by the end of each month, they will receive certain amount of wages. Perhaps, it may look good for us to have a number of subsidiaries, but I think it is just a burden for us. That is why we prefer to outsource human capital.

I have recently suggested to our Board of Directors to slow down the impact of subsidiaries although we have our own contracting firm, i.e. Sri Tunas Sdn Bhd and the trading house, Tunas Edar Sdn Bhd, because I can foresee the effects to be on the negative side once we depend too much on these subsidiaries. At the moment, the subsidiaries are mainly dealing with the internal business of the cooperative. Therefore, the financial management of the housing cooperative should be right and transparent.

19.11 Do You Mind Sharing with Us the Minimum Profit Recorded by This Housing Cooperative?

We normally obtain about 35% profits for each project that will be put into the cooperative's coffers. Let us say last year we managed to get RM3 million profit; 10% of the amount will be deposited into our welfare treasure chest. We also have festive chest and special endowment coffer. Our members enjoy medical insurance up to RM20,000. I am proud to mention here that the bonus for staff is usually 2 months.

19.12 What Do You Think About Branding Strategy? Is It Necessary for Housing Developers?

When we mentioned Tunas Muda in this area, people have confidence in us. There are no abandoned projects under our name. In terms of quality, we may not be the best, but it is still comparable with others. So, in that sense of branding, we are doing quite well. In fact, other cooperatives utilised our name to represent them in developing and selling their housing products. For instance, in Baling (Kedah), the Koperasi Pasar Melayu Baling Berhad developed their 12-acre land but used our name instead. The condition is to assure that the project will be successfully completed. Buyers will not be worried with delayed and abandoned project. In this case, we also act as a mentor to other cooperatives who wish to delve into the housing industry. We help them to get the land and design approval, seek respectable contractors as well as show them how to manage every aspect of housing construction. Hopefully, for the next jobs, they are capable of accomplishing the projects in their own capacity.

19.13 In Terms of Marketing, Are There Any Strategies to Boost the Market?

We have our own website. But we also possess databank that consists of potential buyers since people are coming to register for our new project launches. That aside, we also do marketing through Malaysia Cooperative Societies Commission (SKM) and ANGKASA[1] by participating in their activities to promote our products.

[1]ANGKASA (acronym for Angkatan Koperasi Kebangsaan Malaysia Berhad) is an APEX cooperative, recognised by the government as the National Cooperative Movement of Malaysia.

19.14 How About the Aspect of Financial Management?

We have to be prudent in our expenditure. I always emphasise on the benefits and welfare of our members. They should get what they are supposed to have since the main purpose of this cooperative is to help the members. As mentioned earlier, we have three coffers to assist our members in terms of financial capability.

In view of our financial status, we have not had the need to apply for any bank loans so far. We have cash, but we need to supervise our cash flow wisely. Nevertheless, since we plan to grow bigger and move outside Penang, we are expecting to apply for bank facilities very soon, especially as we intend to buy new lands that may warrant us to make use of such facilities in the future.

19.15 If We Could Refer to Other Fellow Players in Kuala Lumpur, They Have Moved Forward by Forming Good Relationship with the Customers After the Delivery Period. What Is Your Opinion on This Situation?

We have formed a strategic partnership with one mortgage company. What we do is, at first, we will contact our buyers. For those who are interested to reduce their mortgage and bank loans, they may do so by liaising with this appointed firm. The firm will consult buyers on the current circumstances of their financial ability either to make the loan or mortgage their assets. It is more flexible. Thus, the interests imposed can be reduced substantially. That is how we help our homebuyers and maintain good relationships with them. It is our hope that the buyers will know that we will always monitor them in a positive way, so that to some extents, their relatives or children will buy our properties in the future. Soon, we will act as a consultant to these buyers and be able to advise them on this matter too.

19.16 With Respect to Management, What Do You Think Makes This Cooperative Perform Better Compared to Other Cooperatives?

From my viewpoint, the Board of Directors should not totally leave the managerial parts to the appointed staff; they must monitor and coordinate the processes and procedures. The Chairman ought to know and take note on every issue. Fortunately, this is my field, and I have experience in dealing with construction, so they certainly cannot cheat on me. Another thing, the work environment should be conducive, with little or no stress. Our board members should also be well mannered and portray positive attributes. Trust among the Board of Directors should be strong. The Chairman must not make decisions without the consent of the Board of Directors. Some administrative procedures such as the issuance of notice for calling of meetings must be improved and streamlined.

Issues requiring major decisions should be forwarded to the Board. The voice of the Board must be unanimous and clear to gain confidence from the staff. In the construction industry, decisions should be made as fast as possible. Sometimes, the connection with other stakeholders hinders the movement of cooperatives as well. Again, your connection must be great especially with local authorities to expedite the approvals because to succeed in this field without the consent from state government, it is impossible to get started. This situation should be tackled wisely. That aside, other cooperatives seem to be lack of capital. As for us, we have ample backup capital and land bank to advance forward. Based on my observation, those incapable of sustaining are always involved in mismanagement especially in cash flow and finance.

Besides that, the land bank is getting scarce, and the price is no longer reasonable especially in Penang. For example, from the valuation made, it can only be priced to RM20.00 per feet, while the offered price is way beyond that value. This occurs due to speculative actions. If we still proceed to buy, our board members will lodge various queries. That is our limitation.

That is why we encourage the spirit of collective decision among board members. Any new suggestions from members should go through the Secretary and will undergo thorough discussion between the Board of Directors. In fact, we also promote the value of teamwork to our staff. Monthly meetings are held to monitor the progress and growth of business.

19.17 How About the Implementation of Goods and Services Tax (GST)? Please Give Your Opinion on This

Actually, housing prices are directly proportional with the demand. If people think the house is good, I do not think GST can hamper them to buy. For me, the housing price in Malaysia is escalating due to the existence of steady demand, together with some elements of speculation that spurs the continuous increment.

19.18 In Terms of Your Own Personality, What Are the Traits That You Deem Have Affected the Performance of This Organisation?

Perhaps, my previous post as the former people's elected representative has garnered respect from the staff. It does help, for they seem to respect me even more. The second one is based on my qualification. With my background as a civil engineer, they tend to listen, and my instruction is highly considered by them. As the Chairman, I need to know everything, especially the technical side. I believe these are the elements that gained me esteem from both staff and board members.

19.19 Penang Is Currently Administered by the Opposition Party, Do You Find Any Related Challenges that Hinder the Progress of Your Organisation?

As I mentioned before, we are collaborating with several private agencies to develop our land. Since Penang is now under the Opposition Party, we are using these partners to represent us for project issues. They will act as a mediator between us and the state government in order to help us solve any arising matters. This situation was encountered by other government-linked companies such as JKP Sdn Bhd and PERDA (Penang Regional Development Authority) that faced some difficulties with new arrangements of the state government that seemed to be against us. Fortunately, we have no problem at all, although we did encounter some managerial and technical delays that are deemed common everywhere in Malaysia.

19.20 What Types of Approach Were Employed to Implement the Corporate Social Responsibility (CSR) Towards the Public?

Our contribution on the provision of low-cost houses has exceeded 2000 units. We merely subsidised this type of houses, as it consumes half of the entire cost, just to build one house. I think that is our main CSR to the community, apart from zakat (tithe) issuance. For our members, free houses were given to them. The first allocation involved landed property that consisted of bungalows, semi-Ds and terrace houses. For the second allocation, the members will get apartments worth of RM350,000 each. As for the third allocation, it is still being planned. I believe they have enjoyed and reaped the rewards extensively.

19.21 For the Time Being, Any Other Diversifications that the Cooperative Wishes to Dabble in?

We actually want to specialise in this realm although there are some proposals from the internal members for us to diversify, but I prefer this cooperative to focus only on the housing industry because once we diversify, our financial and human resources will change and be hard to supervise. Despite that, we bought new lands outside Penang such as in Sungai Petani for 180 acres, Kulim at about 60 acres and few others in Port Dickson that cost us about RM60 million. We are in favour to buy assets for future development and investment. We will certainly continue with this step to ensure our survivability in the property sector.

19.22 How About the Competition from Other Private Developers in Penang?

The competition inherently exists and there are many. But with our brand, we have managed to compete and survive. For your information, non-Bumiputera buyers are the majority purchasers of our houses. Our products are good and relatively reasonable in price. In addition, and as I have stressed, our reputation is good with no abandoned projects in our history.

19.23 KTMSAB Is Known as the Top Housing Cooperative in Malaysia. Are There Any Other Awards Obtained Within the Past Few Years?

Most of the awards are for national recognition only. For overall position in 2014, we are currently at number three of all cooperatives in Malaysia based on the index issued by the Malaysia Cooperative Societies Commission. The criteria used are based on turnover, no rule violations, etc. Also, we are proud that other housing cooperatives are trying to emulate our strategy and approach like one housing cooperative in Johor that follows our conceptual framework in dealing with the housing business.

Part IV
State-Owned Enterprises/Government-Linked Companies

Chapter 20
Darulaman Realty Sdn Bhd

Keywords Darulaman Realty Sdn Bhd • Zainudin Zainul • Bandar Darulaman • Township developer • Cost control • Risk minimisation

Darulaman Realty Sdn Bhd (DRSB) is a wholly-owned subsidiary of Bina Darulaman Berhad, a government-linked company listed on the main board of Bursa Malaysia, and is 54.49% owned by the Kedah State Government via Perbadanan Kemajuan Negeri Kedah (Kedah State Development Corporation) – PKNK. Established in 1983, Darulaman Realty Sdn Bhd is a household name in Kedah for township development. The company purchased about 2200 acres of estate land planted with rubber trees from Kundong Tanjung Pauh Company Berhad. Subsequently, it pioneered Bandar Darulaman in Jitra as a self-contained satellite town of Alor Setar in 1984. Prior to that, Darulaman Realty Sdn Bhd was also involved in the development of Universiti Utara Malaysia (UUM) with the first phase of construction completed by June 1984.

Today, Bandar Darulaman is one of the most established township in Kedah, which includes affordable houses, lifestyle living, commercial development, a water-themed park and golf resort with all amenities just a stone's throw away. In 1995, Darulaman Realty Sdn Bhd purchased 555.3 acres of land in Sungai Petani (it was tranformed into a project called Darulaman Perdana) and 1100 acres of land in Kuala Ketil (it was tranformed into a project called Darulaman Utama). Many other new developments and townships are in the pipeline to ensure sustainability of contribution. Darulaman Realty Sdn Bhd holds 98.2% equity in its subsidiary company, Darulaman Golf Resort Berhad (DGRB), while the balance is in the hands of the public. DGRB owns an 18-hole golf course which is managed by Darulaman Golf and Country Club (DGCC). Set within 189.5 acres of land, the golf club which has been dubbed as the 'Pride of the North' offers premier golfing facilities to members beside other recreation facilities such as swimming pool, badminton courts, gymnasium, conference facilities, food and beverage outlet, children's playground and chalets.

Realising the motto of *We Turn Dreams Into Reality*, the company's total assets stand at approximately of RM150 million while having a team of 40 employees contributing to its strong growth. Since operation, this homebuilder has sold approximately over 10,000 units while achieving high-value appreciation in its develop-

M. Jaafar et al., *Business Sustainability Model for Malaysian Housing Developers*, DOI 10.1007/978-981-10-5266-8_20

ments. More recently, DRSB has introduced lifestyle concepts for its property projects to cater local needs that yearn for contemporary features. Currently, this firm is led by an experienced General Manager, Mr. Zainudin Zainul.

20.1 Mr. Zainudin, Could You Please Briefly Describe Yourself? How Did You Manage to Join This Company?

I was born in Alor Setar, Kedah. I acquired my Diploma in Marketing from the Universiti Teknologi MARA (previously known as MARA Institute of Technology) in 1986 and degree in the same course in 1993. I joined Darulaman Realty Sdn Bhd in 1999. Right after my graduation, I did a lot of things to earn money, like selling burger and satay as well as becoming a painter. This was because of the recession period during that time. I used to go to Kuala Lumpur to find jobs but finally heard about a vacancy at Darulaman Realty as a posting clerk. My job scope was more on recording all in-and-out stuff. Darulaman Realty was expanding at that time, and many vacancies were opened and had to be filled in immediately. I slowly climbed up the corporate ladder, from clerk to supervisor and then executive, manager and finally my latest position as the General Manager for Darulaman Realty. I am proud of what I have gone through, because not many people are loyal to one company, as other diploma holders tend to jump from one company to another that offers better remuneration. During that time, diploma holders were considered valuable resources to employers.

20.2 What Is the Status of Darulaman Realty Sdn Bhd?

Darulaman Realty Sdn Bhd is a state-owned enterprise, founded during the tenure of Dato' Seri Syed Nahar bin Tun Syed Sheh Shahabudin who was Kedah's Menteri Besar in the early 1980s. Now, it is part of Bina Darulaman Berhad Group of Companies.

20.3 Do Kedah Holdings Sdn Bhd and Darulaman Realty Sdn Bhd Represent One Entity?

Actually they are not, as both are separate entities. Kedah Holdings Sdn Bhd is meant for commercial development, while Darulaman Realty Sdn Bhd is focused more on developing residential estates. There was a time when the management of Bina Darulaman Berhad (BDB) wanted to be a public-listed company. Among the needed exercises to be implemented by the company include the sufficing of land bank accumulation. Management then decided to acquire 1100 acres of land bank at

Kuala Ketil (Darulaman Utama) and 400 acres at Sungai Petani (Darulaman Perdana) where both land banks were given to Darulaman Realty Sdn Bhd. The land bank at Sungai Petani actually belonged to Perbadanan Kemajuan Negeri Kedah (PKNK), but it was injected to Darulaman Realty as part of the proposed listing exercise. Moreover, Darulaman Realty Sdn Bhd is also responsible for local economic development. That is why we develop places such as Baling and Jitra which are considered rural and suburban areas, geographically.

20.4 We Heard That Darulaman Realty Sdn Bhd Had Acquired New Land Bank at Pokok Sena. Is There Any Development to Be Planned There?

That land belonged to the state government, and we have been entrusted to develop the area to become the new satellite town for Alor Setar. However, we are having difficulties to develop that area, with the main problem involving the provision of basic infrastructure facilities such as electricity and water. To develop a small-scale residential area, we do not need such facilities, but we are planning to develop a large-scale settlement. Syarikat Air Darul Aman (SADA)[1] has endorsed this problem and told us that the only source for water is from Kubang Pasu, which is relatively far from Pokok Sena. The other problem is regarding the mechanism to discharge the water. Most of the rivers there belonged to MADA (Muda Agricultural Development Authority) and declared under agricultural land, so we have to find rivers that are located far away from the residential area, and for sure, it will incur more cost to us. Besides, the status there is leasehold, not freehold. It will thus be difficult for us to sell leasehold properties in the rural area. Therefore, the project is still 'on hold' until more research is conducted.

20.5 Has the Change in the Political Landscape of Kedah Affected the Direction of the Company?

Nothing much has changed. For the previous government, we have a bit of difficulty in dealing with the federal government since they shared different missions and ideologies. Nonetheless, when the federal government managed to get back this state from the opposition during the recent election, the situation is slightly similar since the business is still running as usual. The state will always enforce their power to implement rules and regulations, but generally we do not receive special

[1]The official name of SADA is Syarikat Air Darul Aman Sdn Bhd, a public water operator in Kedah.

treatments in such matters. If we were given priority when dealing with other government agencies, it would obviously be a different story.

20.6 Are You the One who Makes Decisions in This Company?

Yes, I am. But there is a limit to my decision-making powers.

20.7 What Attributes Have Made You Successful?

You have to be dedicated and loyal to your company. Being eager to know and learn new things is among the traits that should be imbued. Other than that, any additional values such as interest in marketing, especially in the housing business, will definitely help.

20.8 Do You Have Any Formal Property Education?

I have no formal education in properties. My knowledge on property depends on seminars and courses that I had attended and through working experience that I have undergone.

20.9 Is There Any Future Plan for Darulaman Realty to Acquire More Land Banks?

We are expecting PKNK to inject more assets to Bina Darulaman Berhad (BDB) in the form of land banks. This is because we are known as a township developer, where we need huge parcels of land to be developed. The past year has been a challenging one, but we believe we will be able to overcome the economic turbulence by focusing on our core competencies. As our forte has always been township development, we will focus on this key strength and the differentiation of our township from others. We attribute our success to our strong belief and guiding principle which is to deliver quality properties in a timely manner to our customers. As such, our motto *We Turn Dreams Into Reality* signifies our continuous endeavour to surpass our customers' expectations through understanding and anticipating their lifestyle needs and aspirations. In the coming year, the company will continue to be one of the main players in Kedah's property sector. In line with the measures implemented by the government to spur the property sector, coupled with unsold stocks; ongoing projects and new project launches in Bandar Darulaman and Darulaman Perdana, with gross development value of RM109 million; and our capability to secure new land area of 410 acres for new development known as Bandar Tunku Puteri in Pokok Sena, the outlook of the company for the coming years appears promising.

20.10 Is There Any Specific Risk Encountered by This Company That You Can Share with Us?

Our main risk occurs when we cannot deliver the products to our purchasers. They will sue us, and it will indirectly affect the reputation of the state government. Other potential risk is when the landscape of politics keeps changing. By practice, the ruling party will appoint their own people to lead the state-owned enterprises, and our concern lies on the personnel matters that the workers may lack the necessary experience or are not business-minded, which will consequently affect the overall performance of the company. Lastly, there is the issue of bureaucracy. Even though we are a government-linked company, we still face many problems in dealing with other government agencies. Sometimes, for business affairs that require only 14 days, it can be delayed up to 2 months to be settled.

20.11 How Does Darulaman Realty Sdn Bhd Market Its Housing Products?

We are entering the digital world that emphasises on science and technology. Previously, we depend very much on advertisements in newspaper, but now, all information can be obtained through your fingertips. *Mudah.my* is one of the examples of our marketing tools. Other than that, we also gather database of our potential buyers and will personally text them to attend our project launch. On top of that, I still believe that the best means is the recommendation from our buyers via word-of-mouth marketing.

20.12 What Are Your Product Specialties? What Are the Differences Between Yours and Others?

Our specialty depends on three elements, i.e. strategic location, competitive price and quality assurance. However, the company remains watchful of the internal and external environment which could affect the company's potential such as delay in construction, absence of marketing plans, rising building material costs, etc. With this in mind, we are constantly reviewing our project's feasibility and emphasising on product innovation and value engineering to ensure that we are constantly ahead of the market. Our strength in cost control is also another contributing factor towards facing the challenges in the coming years. The extensive and diversified business approach in terms of both products and geographical mix will enable sustainability of earnings in the future years, with all new developments available at its disposal.

20.13 So, How Do You Make Your Products Differ from Others? What Innovative Ideas Do You Implement in Your Development Projects?

Our innovation is solely based on the needs of purchasers. People around here are concerned more on the size of the house as they prefer bigger houses. Purchasers' preference nowadays is to have a modern concept, so we try to implement that idea in every project we develop. But the main problem here is no matter how perfect the design, purchasers will always want to renovate their houses. Although there is a law to prevent them from renovating their houses, without considering other factors like drainage system that may cause flooding, the enforcement is still not there yet. We should follow Australia where the residents should acquire permission from the municipal before doing any renovations, and by doing so, it can be guaranteed that the residential areas will look presentable and pleasant.

20.14 Do You Follow the Statistics of Purchasers' Composition, Based on Race?

The state government policy is 70:30, where 70% is open for all and the remaining 30% is strictly for Malays. In Kedah, Malays also include Siamese and Arabs. But, in reality, we sell more to Malays. This scenario is due to the fact that most of the people in Kedah are Malays and it is also due to political reasons.

20.15 What Is Your Comment on Build-Then-Sell (BTS) Concept that the Government Is Trying to Impose?

The application of build-then-sell (BTS) system is a good initiative, obviously for the government and the public, as it may avoid abandoned projects. As for developers, we have to transfer the costs to our selling prices that will impinge on the buyers. Normally, for one development project, it takes 2 years before delivery, and if within that period, potential buyers think the design is outdated, they will not be interested to buy. Hence, housing developers have to bear this risk. Eventually, developers have to keep giving discounts to ensure their units are sold out.

20.16 How Do External Factors Affect Your Company's Performance?

Implementation of the GST will result in price hiking. Also, the base lending rate (BLR) issued by the Central Bank of Malaysia will influence the buyers' trend. Usually, we would advise potential buyers to double check with commercial banks on their loan status. We do not want them to become victims of what they borrow. Due to these factors, I anticipate that this industry is now at its stagnant stage.

20.17 With That, What Is Your Opinion on the Current Housing Development Scenario in Malaysia?

In my opinion, supply and demand are still there. The only problem is the lack of purchasing power among youngsters. Speculation activities have increased the price of houses, and the government is doing their best to curb this problem as they impose a relative new law that buyers cannot simply buy and then resell the properties with higher price within a year, or they have to pay 30% capital gains tax. This is the reason for the price escalation, making the younger generation nowadays cannot afford to own a house. At Darulaman Realty Sdn Bhd, we are looking seriously into this matter and trying not to overprice our properties.

20.18 Can You Share with Us the Working Culture that You Inculcate at Darulaman Realty Sdn Bhd?

I try to encourage my staff to deliver good products with after-sale services. We focus on staff attitudes and make sure they know where our company is heading for. We need to take care of customers' satisfaction as we believe in their word-of-mouth recommendations. We keep our promises to our customers because they will invariably spread the positive news of our products to their relatives and friends.

20.19 Do You Have Any Mentor Who You Perceived Have Steered You Towards Success? Who Is He?

My previous boss, Dato' Jahubar Sathik Abdul Razak, inspired me a lot. He is the one who formed Bina Darulaman Berhad. He is a disciplined person, well managed in terms of time, and walk-the-talk person. His favourite quote is 'the cost is within our control, so try to minimise the risk'. He memorises almost everything and is very knowledgeable in engineering and legal terms.

20.20 Could You Please Provide Any Advice to New Property Developers Who Wish to Delve into This Industry?

In spite of everything, before you dabble in this business, make sure you have adequate knowledge about the industry. You have to understand the market needs. It is important for you to understand the people trends since people nowadays like to buy a concept rather than a house itself. They prefer to buy a house that can offer them a newfangled and comfortable concept.

Chapter 21
Binaraya PKINK Sdn Bhd

Keywords Binaraya PKINK Sdn Bhd • Mohamad Nasriff Daud • State-owned enterprise • Islamic norms • Affordable houses • Kelantanese

Binaraya PKINK Sdn Bhd is a state-owned enterprise in Kelantan that is involved in housing development. It was established on 22 June 1981 with a paid-up capital of RM2.26 million with three major shareholders, i.e. Permodalan Kelantan Berhad (51%), Perbadanan Kemajuan Iktisad Negeri Kelantan (38%) and Usahawan Bumiputera (11%). As the main player in providing houses for people in Kelantan, the vision of Binaraya PKINK Sdn Bhd is to be the main provider of affordable homes in the state.

Aside from Binaraya PKINK Sdn Bhd as an entity, there are other two associate companies, i.e. Profil Impian Sdn Bhd (PISB) and Sinaran Pantai Timur Consortium Berhad (SPTCB), which also operate as small-scale developers under the helm of Binaraya PKINK Sdn Bhd. All these three companies are managed by a General Manager (GM) who is accountable and directly reports to the Board of Directors. The position of GM is on a contract basis for 2 years. Effective from January 2014, the company was led by Mohamad Nasriff bin Daud. All the three companies were managed by him through sharing the same management and resources.

Since 1983–2010, the group of companies have completed 2958 units of houses with sales value of RM202.94 million. The most popular type of houses are terrace houses (single and double storey), bungalows and shop houses. The largest development implemented by Binaraya PKINK Sdn Bhd is Desa Kemumin which is located in Pengkalan Chepa with a total sales value of RM88.45 million that consists of 1634 housing units. Currently, the major foci of the group are bungalows and single-storey terraces in Jeli, Kuala Krai, Kota Bharu and Tanah Merah.

21.1 Can You Briefly Share with Us Your Background?

I graduated from Ohio University, USA, in industrial engineering in 1984. After I completed my studies, I worked with Telekom Malaysia and MK Furniture Sdn Bhd before I decided to manage the first Malay McDonald's franchise business in the

M. Jaafar et al., *Business Sustainability Model for Malaysian Housing Developers*, DOI 10.1007/978-981-10-5266-8_21

east coast of Peninsular Malaysia. That was the idea from my close friend who offered me to run the business in partnership. This offer gave me an opportunity to continue my career in my birth state. I would say that through this franchise business that started in 1993, I gained valuable experience since I learnt a lot from McDonald's especially on marketing because it is an international company.

To run the McDonald's business, I underwent a special bachelor degree in Hamburgerology from Hamburger University, USA. I took the course and graduated from Hamburger University within a month. The business ran well until my friend decided to diversify into property investment in Kelantan. Unfortunately, to venture in the property business during the 1987 recession was a not a good step. Major investors withdrew from the project. The failure in property investment affected our McDonald's franchise business. Eventually, lack of cash forced us to hand over back the franchise to the McDonald's of Malaysia.

21.2 Where Did You Gain Experience in Housing Development?

I gained my first experience in housing development when I started to work with E'sha Bladdar (M) Sdn Bhd from 1997 to 2004. This is an interior design company which produced houses for segmented high-end market in Kelantan. We built houses for specific clients who chose to have a special design. Interestingly, many Kelantanese prefer to have an interior designer to design their houses, not an architect. These 7 years of experience with the company allowed me to be involved further with construction and renovation for housing and office projects. Subsequently, my second experience was when I worked with PKINK's (Kelantan State Economic Development Corporation) construction arm, known as Mega Megah Sdn Bhd, which is 100% owned by the state corporation. Coincidentally, Mega Megah is also involved in property development, interior design and building contracts. We were only involved with small-scale housing projects with development of up to 20 units of houses. Thus, my prior experience in housing development for almost 10 years has convinced the Board of Directors of Binaraya PKINK Sdn Bhd to hire me as the GM of this company.

21.3 Could You Please Narrate a Bit About This Company and Your Role?

Kelantan is an interesting state that possesses unique characteristics compared to other states in Peninsular Malaysia. Politically, this state has been ruled by the opposition party since 1990. As a state-owned enterprise, Binaraya PKINK Sdn Bhd is owned by the state government. Together with two associate companies, i.e.

Profil Impian Sdn Bhd (PISB) and Sinaran Pantai Timur Consortium Berhad (SPTCB), I am responsible for the ups and downs of this business group. As the leading company, Binaraya PKINK Sdn Bhd handles all the company management and will charge certain management fees based on percentage of managing SPTCB and PISB. Since these three companies pursue similar corporate vision and mission, my additional role is to evaluate potential lands and present it to the Board. The Board will then decide which company among the three should buy and develop the land. The decision will rely on the financial capacity of each company.

21.4 What Is the Main Objective of the Foundation of Binaraya PKINK Sdn Bhd?

We are incorporated to accelerate the strides and efforts of providing homes for the state populace and subsequently be a leading premier property developer in Kelantan. Also, our central onus for the state is to assist poor Kelantanese to own a house. I would love to share with you that we have lent a hand to the state government to build a quantity of houses for the underprivileged and selected folks, with a value of RM28,000 each in 2013.

21.5 How Do You Position This Firm in Comparison with Other Government-Linked Companies?

Our parent company, Perbadanan Kemajuan Iktisad Negeri Kelantan (Kelantan State Economic Development Corporation), has a number of companies which run similar businesses, such as Kelkon Sdn Bhd and SPP Development Sdn Bhd. They are our sister companies. Our subsidiaries are also involved in the same field, either construction or property development. I would say that Perbadanan Kemajuan Iktisad Negeri Kelantan is in the process of structuring the property segment in order to become the major player of the industry in this state. We are striving to beef up our business in order to be recognised as the leading government-link company (GLC) in property development. For your information, the role of each company can be separated by the type of development such as landed property, high-rise building and affordable house.

21.6 Is There Any Specific Target or Vision Set by the Board?

As a General Manager, I need to make a viable plan on how Binaraya can make RM3 million in a year. We have diverse types of land banks, scattered around this state, although the majority shares belonged to our parent company. At the end of the day, we need to pay dividends and shares to many parties. Since we deal a lot with state administration, we also need to allot quite a high allocation of our profit for corporate social responsibility.

21.7 What Is the Current Performance of This Company?

As it stands currently, we have developed houses in Jeli, Kuala Krai, Machang and Tanah Merah. Honestly speaking, land prices in Kelantan are among the highest in Malaysia, for example, in Kota Bharu, it can reach RM30.00 per square feet. For a bungalow with 1700 square feet, property developers are selling it for RM400,000 and higher. In Kelantan, the demand is fairly satisfactory, and price is not a crucial element for the buyers especially in Kota Bharu. We have a good brand, so we do not have problems in selling houses. People are confident with our performance. However, in Kelantan, homebuyers still prefer landed properties. To do high-rise property development requires strong financial backup. We have many land banks with the majority located in Kota Bharu, Machang, Tanah Merah, Kuala Krai and Jeli.

Ironically, although the general perception that the state is poor, with the annual growth being much lower compared to advanced states like Selangor and Johor, the buying capacity of Kelantanese towards housing products is high. With the increase in population, the demand for housing will be significantly responsive and is expected to boost further.

21.8 Do You Mind to Reveal the Total of Houses Built and Barriers that Hampered Your Company to Provide Affordable Homes to the People of Kelantan?

Since 1990, we have completed almost 10,000 units of houses in the whole state. We are in the middle of developing another 300 units this year. The earliest project by Binaraya PKINK Sdn Bhd was delivered somewhere in 1983, located in Desa Darulnaim, Pasir Tumboh. Our associate company, Profil Impian Sdn Bhd, also involves in many housing development activities. Its flagship project named Taman Kurnia Jaya was initiated in 1995, comprising 352 units of low-cost houses and 156 units of middle-cost homes that cost us of more than RM10 million. Due to high demand from the public, we also developed 58 units of bungalow in Kampung Jelatok, Tanah Merah. Each house was sold at RM160,000 in 2007, and the present value is recorded at as high as RM250,000.

Needless to say, housing demand among Kelantanese is still unwavering especially from the young generations who have just started working. Nevertheless, many property developers in Kelantan are keen to build middle-cost and high-cost houses that may give impact to low-income earners. That is why you could see that majority of Kelantanese prefer to self-build their houses as their family also typically owns few parcels of land. I personally observe that the low-cost houses can only be built if the government enforces a ceiling price for building materials and the project is executed on the state-owned land. Otherwise, it would be difficult for private homebuilders to develop low-cost residences.

21.9 What Is Your Vision in Managing This Company?

I need to manage staff based on their key performance indicator (KPI). Previously, there were no systems used to assess staff performance. We need to reward staff based on their performance. I have vast experience in managing business with McDonald's and was involved in human resource and accounting restructuring in Telekom before. I relooked on their job scope and effective working time. For the firm's strategy, we are now trying to find land banks outside the urban area with 4–5 acres in size. However, the infrastructure cost is getting higher as it can go up to 20% of the overall costs. The main challenge of this industry is to determine the right selling price to avoid failure as affordability is limited.

21.10 Since You Hold an Important Post in a State-Owned Enterprise, Can You Share Your Future Vision for Kelantan?

I strongly believe that Kelantanese still need houses. I hope that the state government can focus on developing houses for people because we can easily obtain the resources such as sand, cement and steel. We need to go for a bigger-scale development, but we need foreigners to invest in various industries in Kelantan. Government-linked companies like us may then focus to create new sustainable townships. To begin with, I would like to propose a new Islamic township. We can provide hall and enough facilities for community activities, and they can gather every Friday for religious activities.

At present, we are operating the business like other private housing developers, not as a state agency. Perhaps, Binaraya PKINK Sdn Bhd should focus more on bigger developments with projects of 200 to 300 hectares. So far, all of our houses are sold out. We do not face problems in selling houses since the salient advantage here is, although house prices in Kelantan are rather equivalent to Kuala Lumpur, people here, however, do not spend their money on entertainment, making their disposable income very high.

21.11 What Is the Strength of This Company?

Until now, Binaraya PKINK Sdn Bhd has no abandoned projects. All our projects are completed on time, and financial assistance from the banks are easily obtained. Again, Kelantan is not similar to Kuala Lumpur. There are no specific demands in terms of innovations, and emphasis on green aspect in our design is considerably irrelevant. Whatever we erect, our project will be sold out. For each project, we put a benchmark of 20% for profit margin regardless of the project location.

21.12 As a Company Under the State Government, What Are the Limitations for This Company in Terms of Performance?

PKINK is owned by the state of Kelantan, whereas the majority of directors or shareholders are politicians. Sometimes, politicians and professionals have different ideas. Besides that, the contract position is only valid for 2 years for very important posts such as GM and Chief Executive of the company which inextricably limits the power to establish changes. The working culture here is different, as they are government servants and are difficult to push. I need to set individual KPIs and reward those who perform well. I do anticipate, however, that with the current management structure, our group will be able to become a good corporate government-linked company (GLC) in helping the state to provide affordable homes for Kelantanese.

21.13 Do You Have Any Specific Characteristics that You Think Have Contributed to Your Success? Next, Please Share Some Advice for Novice Homebuilders

Again, my Islamic upbringing seems to help me a lot in paving the journey of my life. Although the odds were stacked against me, I must say that discipline, a positive attitude, specialisation and a strong mindset pushed me forward. You need to love what you are doing. Do not let the circumstances around you determine your mindset. For nascent homebuilders, please be reminded that it was very easy to start the business a few years back. It may still be possible now, but you are going to need a lot more capital and patience these days. As the property market is nowhere nearly as active as it was in the early mid-2000s, banks and other financiers are reluctant to lend even to successful and experienced developers. As a new developer, you are really going to struggle.

21.14 Lastly, What Is Your Personal Vision?

This post is only a contract basis. I grew up as a son of a politician, and my mother is a religious teacher who provided me with good guidance for children especially in terms of Islamic teachings. This religious approach attracted me to give an Islamic-based education for my kids. Now, I am learning Arabic, and my future interest is to teach professional people to understand the Holy Quran by learning Arabic. Also, I hope that with my religious knowledge, I would be able to spur superior performance of my subordinates as additional values for them to work harder and smarter as demanded by Islamic foundations.

Chapter 22
Perbadanan Kemajuan Negeri Selangor

Keywords Perbadanan Kemajuan Negeri Selangor • Azlan Md Alifiah • Parastatal organisation • Customer orientation • Paradigm shift • Social responsibility

Formed on 1 August 1964, the Selangor State Development Corporation or Perbadanan Kemajuan Negeri Selangor (PKNS) may be a veteran in the property development industry, but the organisation is getting agile to compete with other real estate actors in order to contribute for national economic growth. Generally, PKNS is the biggest state-owned enterprise in Malaysia, with equity shareholder of RM5 billion. It is a large and viable organisation. The recorded group profit in 2010 was RM400 million. In terms of development concept, PKNS is still the main developer for township development in Selangor, having numerous projects in Petaling Jaya, Kelana Jaya, Ampang, Shah Alam and other parts of the state. From its initial challenges in developing a township to achieving a record profit of RM420 million in 2011, Perbadanan Kemajuan Negeri Selangor (PKNS) has come a long way since it first started. The name has more often, in the public's eye, been associated with social housing – providing affordable homes for the population in the state which in truth was its initial objective. But, there is more than meets the eye for the Selangor state government's development arm in recent years as it has ventured into asset management and hospitality industry, to name a few.

Azlan Md Alifiah is the sixth General Manager of PKNS since February 2014. He was born in Ipoh, Perak, and is an accountant by training, having graduated from Universiti Teknologi MARA and the Chartered Institute of Management Accountants (UK) in 1991. He has 22 years 'hands-on' experience and exposure in the business and financial operations and day-to-day activities in finance and accounts, investments, insurances, risk management and corporate governance functions either at group level or at the strategic business unit levels. He has held various senior positions in financial and telecommunication firms including Head of Risk, Shariah and Governance at Prudential BSN Takaful Berhad; Vice President, Risk Strategy at Etiqa Insurance and Takaful; and Senior Manager, Finance for Shared Infrastructure at Celcom (M) Sdn Bhd. His first job in 1992, however, was as a Management Account Executive at Petronas Carigali Sdn Bhd. Prior to PKNS, he was appointed as the Chief Financial Officer of Kumpulan Darul Ehsan Berhad (KDEB), the investment arm of the Selangor government in June 2011. He was instrumental in

M. Jaafar et al., *Business Sustainability Model for Malaysian Housing Developers*, DOI 10.1007/978-981-10-5266-8_22

the corporate exercise and privatisation of Kumpulan Hartanah Selangor Berhad duly completed by end of 2013. At present, he is one of the Board of Directors for various subsidiaries and associate companies of PKNS.

22.1 Could You Please Succinctly Expound to Us About Your Organisation and Its Vision on Property Development?

Selangor is brimming with opportunities even in the early 1980s, due to its close proximity with Kuala Lumpur, the capital of Malaysia. Long before countless big players dominated the real estate scene in Selangor, Perbadanan Kemajuan Negeri Selangor (PKNS) has already been stamping its mark as the giant homebuilder in the market. PKNS tapped the potentials by putting their real estate experience and acumen to work by building high-end properties in Selangor at par with other best housing developers anywhere. As the master developer and pioneer of Selangor, many of our products have transpired into success stories today. We are a very active and responsible developer who wants to remain competitive and relevant by trying to put our highest concern on the level of customers' satisfaction. We are also the mastermind behind 11 major integrated townships such as Bangi, Shah Alam and Petaling Jaya. At present, apart from property development, we have diversified into asset management and hospitality sector.

22.2 If You Do Not Mind, Kindly Please Brief Us About PKNS and Its Development Policy?

PKNS itself possesses wide experience of more than 52 years in township development. We are among the pioneering developers in Shah Alam as we have developed almost 10,000 acres of land there. Now, we have about 10,000 acres remaining land, mainly in Bernam Jaya and Kota Puteri, and we are still looking forward to expand and acquire new land banks in this state. To a larger extent, we will continue to develop more townships and try to seek for new yielding investments especially on major high-rise developments such as Datum Jelatek, Datum InCity, Datum EduCity, etc. We are estimating the projected GDV for the next 15 years to be approximately RM12 billion.

Our focus from the inception till now is still on property development. PKNS is striving to become the best state-owned enterprise in Malaysia. In tandem to the national vision to provide sufficient houses for citizenry, we think our roles are pertinent in building community, empowering their life and realising their dreams as well. The idea inevitably coincides with our main vision. The Government of Selangor has embarked 'Selangorku' (My Selangor) project to cater the fundamental needs by erecting low- and medium-cost houses for the state populace. In conjunction with this campaign, we are expecting to build about 7500 units starting from 2015 to 2020 and 30,000 unit of houses by 2030. Based on our own record, we have met the call of government to develop affordable homes, where 1422 units were built at Bukit Botak in 2013 and another 124 units in Bangi.

As a state-owned enterprise, we are willing to shoulder the responsibility to erect affordable houses in this state although it will incur loss to our side. That is why we subsidise those projects with high-end houses such as all the Datum projects. In order to ensure that clear demarcation on the role of PKNS as a state agency, our subsidiary that develop Datum projects will cater on high-income demands, focusing on high-rise properties, while another subsidiary, Worldwide Holdings Berhad, will concentrate on the high-end landed properties. For PKNS itself, our market range is between lower-income group to the middle earners. We have our own brands and companies to undertake different market segments. Consequently, from the profits obtained, we will build affordable homes for the state population as mandated by the government.

22.3 Anyway, What Are Your Generic Responsibilities as a General Manager of PKNS?

As a General Manager, I am entrusted to improve the performance of PKNS specifically on the issues of corporate governance. I joined PKNS when it was in a good form, which differs from my previous experience as I was dealt with uncertain performance of organisations. So, it is more on corporate turnaround, a bit of financial restructuring and eventually expanding the business of PKNS.

Before this, I had joined Kumpulan Darul Ehsan Berhad (KDEB), the investment arm of the Selangor government, in June 2011, and I was among the key persons in structuring Kumpulan Perangsang Selangor Berhad (KPSB)'s disposal of equity interest in Kumpulan Hartanah Selangor Berhad (KHSB) (both KDEB subsidiaries) to KDEB.

22.4 May You Please Share with Us Your Prior Working Experience?

I have a diverse experience in terms of industries, ranging from oil and gas to healthcare to the fast food business to telecommunication before delving into property sector. Prior to PKNS, I was with Kumpulan Darul Ehsan Berhad (KDEB). I was instrumental as part of the management team to turnaround KDEB Group's investments and performances, where I obtained valuable experience encompassing business turnaround management, corporate finance as well as business and financial restructuring exercises. Before KDEB Group, I have held various senior management positions in the financial and telecommunication industry. From 2000 to 2003, I was the Chief Financial Officer of Kumpulan Medi Iman Sdn Bhd and also a Director of A&W Malaysia, Singapore and Thailand under the same organisation of TDM Group. During my tenure there, I was involved in its turnaround management and in the sale of A&W (M) Group to KUB Malaysia Berhad. I started my career with Petronas Carigali from 1992 to 1993 as Management Account Executive, KPFB Holdings Sdn Bhd as an Accountant from 1993 to 1994 in the segment of palm oil industry and Celcom Technology as the Head of Finance from 1994 to 2000. Within that period, I was promoted as Senior Manager Finance, focusing on infrastructure projects to manage the assets and implement the contracts and projects.

To go further, telecommunication has a direct bearing with what I am doing now in PKNS because I was the head of infrastructure group. Thus, I am quite familiar with civil and mechanical works as well as project management while capable to accomplish project management and accounting in regard to coordinating project and dealing with technical matters of engineering and consultancy. Apart from that, I had involved with turnaround operations for two loss-making hospitals. Interestingly, with my capability, one of the hospitals managed to record profitability, while the other one was on the route to be stabilised and make profits. I have also done on the disposal of assets with A&W and taking over the companies, the last one during my tenure with Kumpulan Hartanah Selangor Berhad as we took over a private firm during the general offer.

22.5 What Are the Skills That You Have Acquired Throughout Your Working Hindsight?

I would view the acquired skills through the same lens. Since I am a chartered accountant by background, cost management is one of my specialisation. Analysing skills of revenue, assets and their cost components are my core competencies.

22.6 What Do You Think About Your Personal Strengths That Might Be Useful for PKNS?

My strength lies on asset and finance management as well as corporate and financial turnaround. Hence, my main task here is to utilise my expertise in portfolio management to diversify the income of PKNS into few other segments. We will observe both upstream and downstream activities to realise such goals. Admittedly, about 90% of our income are derived from property development. In economic situation which is normally cyclical in nature, we need to seek new channels of income to maintain the profitability of PKNS. In light with that circumstance, PKNS has embarked a new entity called PKNS Real Estate Corporation (PREC) in 2012.

22.7 Do You Mind to Share with Us the Roles of PREC?

Being a private limited company, PREC will be able to focus on real estate investment and asset management, while PKNS will continue to focus on its core business of property development. As promised, we will maintain ownership of our commercial assets comprising Kompleks PKNS Shah Alam, Kompleks PKNS Bangi, SACC Mall Shah Alam, Menara PKNS Petaling Jaya and Wisma Yakin and will transfer them to PREC. PREC will continuously upgrade the buildings it owns to enhance their asset value and image. For example, RM10 million has been allocated for the maintenance and upgrading of Kompleks PKNS Shah Alam. The establishment of PREC is basically intended to act as an incubator to yield up and manage PKNS's assets in order to attract good profile tenants and retailers. With PREC, we hope to acquire more buildings with upside potential through asset enhancement exercises and to increase the capital value of properties by improving yield. A ‘value creation exercise’ would be carried out periodically to make sure that the assets owned were competitive and relevant to current demands and lifestyles. For the time being, they will bring new tenants to our shopping complexes and apply refurbishment towards the facilities in the buildings. As a result, the demands and responses from the public have been increasing that subsequently assist the business of our tenants. We have inevitably brought new markets and benefits to the tenants by raising the profile of our services. Rather than being passive, waiting for the people to come, PREC has played his role to pull the market to these places.

Obviously, we are responsive to the calls by traders for a suitable location for their businesses that can attract customers, has good amenities, and is well maintained. PREC with the cooperation of PKNS' Entrepreneur Development Division will add more sites in its five commercial assets for Malay and Bumiputera traders to sell their products. For a record, since 2008, over 6000 Bumiputera entrepreneurs have been benefited from our entrepreneur development programmes. I perceive that we as a landlord, we need to add value to help our tenants by attracting market to them. We are also mindful that as most of them are Bumiputera retailers, we wish to ensure that their business will prosper together with our advancement. We create synergy with few other partners to strengthen our desire in retail markets. Going forward, we are planning to expand our venture in retail markets also through De Palma Bistro. At the moment, we have one outlet in SACC Mall. Future prospects may entail the vision to open new outlets at each of our commercial buildings. It is not so much about gaining profits from this type of business as the idea is to develop and facilitate local entrepreneurs.

22.8 Would You Mind to Elaborate on the Entrepreneurship Programme Embarked by PKNS for the Retailers?

For your information, PKNS is planning to open two shopping malls in Datum Jelatek (Kampung Datuk Keramat, Kuala Lumpur) and Taman Selayang Mutiara; in addition, Tower 2 is another iconic building to be built in PJ Sentral, all of which are expected to be in operation between 2018 and 2020. PKNS had selected 200 Bumiputera entrepreneurs to undergo intensive course in various aspects of retailing, with a focus on merchandising and retail concept as well as internet strategy. The two-year stint was expected to enhance the Bumiputera entrepreneurs' competitiveness by placing them at high-end shopping complexes apart from PKNS's properties. For a start, we will place them at Datum Jelatek, and if there are other eligible entrepreneurs, we will put them at our other business complexes. We welcome major shopping mall operators in the country to collaborate with PKNS by placing Bumiputera entrepreneurs in their business premises.

22.9 What Have You Done in KDEB That You Think Might Be Beneficial to PKNS?

When I came in (along with new directors) to the management of KDEB, we have to look at the organisation and its people as we try to optimise and restructure the organisation in order to enhance the business efficiency and effectiveness. We believe that all corporate endeavours start with human parts. Most of the business branches deal with people. We change the management even though it was quite painful for them. The stride is merely for the sake of business improvement. We are different with other organisations because we infuse Islamic principles and values in our daily works. We have appointed Chief Religious Officer in KDEB as we envisage to incorporate Islamic teaching in every business dealings. Then, during our second year tenure, we had begun with the improvement of portfolio management and asset allocation in terms of the balance sheet of the company. We look at our prior investment that produced negative yields and try to execute such businesses into profitable ventures.

I think that those are the reasons that made me selected, i.e. to turn around the company and address the lack of performance. For PKNS, in the next stage, we plan to focus on the investment field with respect to property development. In this sense, we want to be very clear on the product viability and target market. Since we have few other portfolios, we will maintain and effectively manage them to gain positive results. We have basically a specific team to handle these investment strategies. They will advise the management either to continue or divest the investment. Before we do that, we need to be able to identify the goal of the prospect business especially where it will lead you and what are the benefits that will prevail? We must anticipate that within 3–5 years, we will get higher returns from the dividends.

Besides that, we also create businesses that might supply recurring income to us. At that point of time, such income may be attained from PREC and other investments made by us. We should be very tactful in investing strategy, either to decrease the emphasis, dispose the assets or expand the investment growth.

We also attempt to emulate the business model of PNB who owns diverse business segments in its pipeline. Whenever one type of businesses faced difficulties during economic recession, other businesses will be able to cover. As a state-owned enterprise, our current emphasis is still on property development, but we hope to enlarge the pie into other segments of investment as well. After all, we need to be prepared with sudden economic crashes based on forecasted business cycle.

22.10 Since You Were with Other Sectors Before, How Do You Gain New Motivation to Restart Your Career in the Property Development?

By doing the turnaround in KHSB, taking it private, looking at the pitfalls and the problem we had, that really piqued my interest in the property business. Property, if you compare to other types of business, is something that is a need of everyone. There are three things that you need: the clothes that you wear, the food that you eat, and also the roof over your head. So, property is very close to people's heart. Besides that, when we look at the property business, there are things that we can bring in. I have been in the financial industry, also in the civil, mechanical and electrical sector as I found that there is always an element where we can use the best in all these industries to bring to PKNS as a property developer.

22.11 May We Know the Recorded Profit of PKNS for the Past Years?

Profits that we have made are going to be regenerated for our core deliverables. We achieved the highest profit in 2011 as we applied major exercise to clean up our balance sheet in terms of our stock overhang. We managed to accumulate RM400 million. Afterwards, within 2012–2014, our profit was recorded between RM200 million and RM300 million. It can be said that we are undergoing sustainable period of growth. Our gross revenue is closed to RM800 million a year. Within the next few years, we are targeting to have another approximately RM3 billion of GDV from the Datum series for the next 5 years.

22.12 What Are the Current Challenges Faced by PKNS?

The cost of land acquisition and construction materials is hiking, but the framework is still there. It just needs a little tweaking. We will further exercise acquisition on lands to increase our land bank. At present, we have about 10,000 acres of land bank, situated mostly in less-developed parts of Selangor. To enhance our land bank, we are looking forward to discuss with landowners in matured areas such as Petaling Jaya and Shah Alam to redevelop the locations. This approach is called 'zero-land bank strategy' in order to replenish our land supply.

22.13 Accordingly, What Are the Overarching Strategies Employed by PKNS to Remain Competitive?

We have made a valiant effort in the last 10 years to make ourselves more efficient and popular among Selangor people. To be honest, our focus now lies on the project execution. My mandate for the first year is to look for efficient people and effective organisation, while for the second year, I will focus on portfolio balancing in terms of assets, enhancing our yields and stretching our capital efficiency. In the third year, we expect to optimise our profit. We want to have sustainable income and growth. We are thinking to create values on our land and projects.

Sometimes, you cannot develop the projects by your own, which triggers you to joint venture with other housing developers. Otherwise, you will be curtailed your working capacity and own reputation. Besides, we are synergising our properties with current development set by the government. It happens with our projects in Bernam Jaya that we prefer to wait for the best moment to start the official launch. Like our joint-venturing strategy, for example, our collaboration with Sunway Berhad for a project in Kota Damansara has given us a new dimension of township development. Apart from good dividend yield, the innovations portrayed in the projects are tremendous. You may take a look at Sunway Giza; it injects different environment to the scene. All in all, in terms of privatisation and JV, our GDV has amounted to RM1.9 billion, and it can be considered as a success.

Another strategy of PKNS is that it is designed to not only reap profits but return it back to the public. As far as the financial goes, since day one, we managed to create a value of close to about RM5 billion net tangible assets (NTA), as of 31 December 2014, and this NTA is created from the returns we get from the project before we cycle back into acquiring land like Selangor Science Park 2 and also Bernam Jaya. If you look at that perspective, PKNS is holding the wealth for the people in that context, and we are then able to deliver for future generation. So, as far as PKNS is concerned, it has been designed that way. And what we are trying to do now is trying to reap this in an orderly and sustainable manner.

For your information, we have made our foray into Australia too. We are looking to have profitable returns the next few years based on our investment in Perth through our arm, Worldwide Holdings Berhad. We are anticipating opportunities to have projects outside Selangor, but they are still under discussion. It requires negotiation, capacity and stamina to ensure our people to be kept abreast and trained within the challenging times. Also, we are planning to cooperate with hospital operators for us to be their building owner. Selgate Corporation Sdn Bhd, our wholly owned subsidiary company, will undertake the processes.

22.14 Does PKNS Involve in the Provision of Affordable Homes?

Being a state developer, PKNS has multiple roles to accomplish, and one of them is to provide affordable homes for the people. To achieve this goal, PKNS subsidises a major portion of the cost in their developments. In fact, most of our projects are heavily subsidised. For our recent project at Selayang Mutiara, we sold the semi-D houses for RM99,000 when it cost us RM180,000 while the market value is RM350,000. So whoever bought the house, it is like winning a lottery. Personally, PKNS's business model appeals to me because it is a property development company that is not looking for profits only but also at its social responsibilities. I am at a certain age where I believe that everybody needs to return something to the public, to the people at large.

Now, the ceiling price for affordable houses is RM250,000, while the minimum rate is about RM70,000 to RM80,000, opened for the accumulated income of the households of RM4,000 to RM6000 monthly. In Selangor, if you earn less than RM1500 a month, you are considered at the poverty line. For the low-income earners, we will build affordable houses for them, but the allocation process is managed by the Selangor Real Estate and Housing Board (Lembaga Perumahan dan Hartanah Selangor – LPHS). Thus, they must firstly be registered with the board for further evaluation. As part of our corporate social responsibility (CSR), we are responsible to build affordable homes for Selangor's citizenry. In city centre and suburban, we are offering high-rise residences with 3 bedrooms of about 600 square feet for low-cost houses and 850 square feet for affordable homes. If they wish to own landed affordable houses, they may need to commute to rural areas such as Antara Gapi in Hulu Selangor and Bernam Jaya, which is near to Tanjung Malim. They might be far from Kuala Lumpur, but with planned accessibilities such as highway and train, the housing estates are deemed viable and potential.

We find that the people of Selangor have difficulty in buying houses in the state as their prices are above RM500,000 per unit. As such, PKNS has taken the initiative to make available the affordable homes that cost between RM150,000 and RM250,000 while helping to fulfil the state government policy of committing to build more affordable residences. Selangorians who met the criteria and had registered with the LPHS would be given preference in the ownership of houses offered by PKNS. We target to provide at least 7500 affordable homes from 2015 to 2020.

We plan to provide more affordable homes in the next 3 years in line with the state government's development programme. Currently, PKNS has built 1977 affordable houses in Taman Selayang Mutiara, Bandar Baru Bangi, Seksyen U12, Antara Gapi and Kota Puteri, respectively. This social obligation at the end of the day gives benefits to the whole strata of people of the state of Selangor.

22.15 How About the Recent Projects of PKNS and the Strategic Plan for Upcoming Years?

Within the next 5 years, our projects will cover mainly in northern region and central region, with some in the southern region. For example, our affordable housings that we are targeting to construct in 2017 are in Selangor Science Park 2 (in Cyberjaya) with 2000 units by phases. Other than PKNS itself, our subsidiary companies like Worldwide Holdings Bhd and Selaman Sdn Bhd also develop affordable homes. Worldwide has successfully developed between 200 and 300 units of affordable houses in Puncak Bestari, while Selaman undertook affordable homes in Padang Jawa last year.

Another new project is Kota Puteri, which was designed to be a township of choice for the younger generation. It can attract various segments of society to buy houses in Kota Puteri, which is now accessible via several highways. On top of that, landed property in Kota Puteri can still be purchased at between RM350,000 and RM450,000. We are now in discussion with the state government to improve the public transport system in that area.

22.16 Land Bank Is an Indispensable Resource to Housing Developers. How Does Land Bank Might Be Related with Your Notable Projects?

PKNS has two big land banks, especially in Selangor Science Park 2 in Cyberjaya that will start major development by this year. We are now in detail discussion to find an education catalyst because having an educational centre from the pre-school to the primary and secondary and going forward to tertiary would allow us to attract the necessary community. The planning of Selangor Science Park 2 is interesting as PKNS is set to build its new satellite office there, along with shopping malls, plus 2000 units of affordable houses and also three plots of interconnected lakes. This is aligned with the rebranding practice of Selangor Science Park 2 with its new name, Selangor Cyber Valley.

Our second land bank is located in Bernam Jaya that I should describe as the 'higher hanging fruit' which I think is going to mature within the next 5–15 years. Aside from residential properties to be built there, we are also focusing on creating more avenues for business in Bernam Jaya. We felt lucky because Bernam Jaya is connected to PLUS as one of the largest highway concessionaries in Malaysia. We are looking at developing universities and potential maintenance, repair and operations (MRO) too. Hence, it is possible if you want to extend there to become an airport or other MRO facilities. Transportation-wise, Bernam Jaya has the North-South Expressway, and Komuter train has already been running across the development. We are in discussion within the next few years to build a station there. This will allow for the population in Bernam Jaya, either they work in Proton City or in Kuala Lumpur, to use that kind of transportation.

This is already in the pipeline. We are just waiting for the right moment to be able to do it. But, like I said, I am an Asset Manager and when I look at assets, I have to put it into their maturity and yielding profile simultaneously. As far as we are concerned, these land banks can last us for another 20 years. Therefore, on our strategy, we also need to look further at our land banks, and we are actively in discussion with the state and also with private landowners to acquire or to buy new strategic land. Even our biggest subsidiary company, Worldwide Holdings Bhd, is also in the midst of getting more land banks. So, this is our strategy to become sustainable developer, to be able to sustain and give recurring income to the state.

22.17 While We Are Glad to See PKNS Develops Suburban and Rural Areas, Do You Have Any Projects Situated in City Centres?

Urban regeneration project is also one of the areas that PKNS has been embarking upon, and one of them is Datum Jelatek. It is 'part and parcel of social obligation' to the people because most of the time, in certain matured neighbourhood, industry should not be there any longer. It should be moved to new area because those areas are now full of neighbourhood of people. Urban regeneration is something that needs to be done. Otherwise, you will create ghettos social problems and also decays. That is why we will never neglect our urban development, and now it will be handled by our associate company. DatumCorp International (DCI), a fully-owned subsidiary of PKNS will be in charged on all Datum series projects. DCI also has other Datum series that it is working on, and one of them is Datum InCity whose phase one is already under construction and set to be delivered by end of 2017, and it is among the biggest projects in Petaling Jaya with an overall GDV close to RM3.5 billion.

We also have other Datum series in Bangi, Shah Alam and Kota Damansara. So, we are looking at developing all of these in phases. Although we have a lot in our plate, we are still actively looking at adding new land banks, but we are very much attuned to the market trend and market outlook. Therefore, we will be looking closely at the market whether market condition is improving or not.

22.18 We Heard That PKNS is Also Involved in Commercial Development. Please Elaborate on the Projects

We are actually banking on a slew of commercial projects in the Klang Valley to boost our construction revenue. Our latest commercial properties consist of shoplots and factories in Kuang, Antara Gapi, and Rasa in Hulu Selangor, as well as Bandar Sultan Suleiman in Port Klang. Perdana Avenue in Kuang, for instance, offers three levels of shop offices. The development boasts a built-up area of 1952 to 4196 square feet with prices starting from RM450,000. For Perdana Avenue, we expect to gain traction due to its strategic location near to recreational areas such as Greenwood Lakes Country Resort, Blue Lake and Golf Club and residential areas that include exclusive semi-detached houses.

22.19 What Kind of Innovations That Have Been Implemented by PKNS?

Indeed, we have much nicer design as compared to before with more receptive to people's taste along with attractive colours and efficient arrangement. Since we are now a market-driven entity, the designs have moved in parallel with the time. One of the key criteria we look at, when we design our master plan for development, is that we design it in such a way it has a balance. It has the opportunity for people to live, work, play, exercise and enjoy. Perhaps, that is why we have this motto: 'Building Communities, Enriching Life, Realising Dream'. We believe in benchmarking ourselves and our developments on standards and quality ratings as we want to ensure that our projects incorporate the latest practice of eco-sustainability architecture.

For our latest prestige housing project, Hijauan Enklaf which is located in Alam Nusantara, Setia Alam, Selangor, whereas, with your very own private pools and lift, living at Hijauan Enklaf is the ultimate in indulgent living. Hijauan Enklaf features rain water harvesting elements, passive shading control, water-efficient fixtures, environmental and user-friendly materials and finishes as well as a structural design that encourages natural class ventilation.

Besides that, we are eyeing to ensure that our new development projects vie for the Green Building Index (GBI) certification. In fact, our RM170 million headquarters in Section 14, Shah Alam has achieved GBI's highest rating of platinum recently. We hope that all of our future developments will receive a minimum GBI-certified rating. We have injected a host of green features including a rooftop garden, rainwater harvesting system, passive design to promote daylight lighting and solar-powered energy. The corporate office is a low-rise building, and the approach of designing a 'green' building started off by orientating the long façade of the building in the north-south orientation, thus reducing exposure to the direct sunlight. By breaking the long building into a series of smaller blocks, it allows more natural light to penetrate into the building, therefore reducing sole dependence on electricity for internal lighting.

The buildings' energy-efficient design reduces the buildings' energy consumption by 60% compared to a conventional building. In that case, the building complex will be insulated and sun-protected and subsequently utilises natural daylight to offset electric lighting. All electrical and mechanical systems will be energy-efficient, and cooling will be provided using innovative energy-efficient ventilation and cooling techniques. A solar air-conditioning system will contribute to reduce the electricity consumption further. This solar cooling will enable 10% of the air-conditioning in the building to be solar-powered, while the green roof will keep it insulated. On top of that, rainwater harvesting, grey-water recycling and water-efficient fittings will contribute to reduction in potable water usage by at least 60%. The new complex will house the PKNS corporate office with a headcount of 600 employees working in the facility.

22.20 How Does Reputation Affect the Business Success of PKNS?

Branding is an issue for PKNS, as many believe that it would not be competent enough to build quality high-end properties. For your information, PKNS has embarked the branding exercise several years ago. Long before the call of government to employ the concept of 'build-then-sell', we have implemented the concept in our projects. We will only start to sell after 70% of completion for most of our projects. Many buyers were confident of the build-and-sell concept, and this has been helpful for PKNS property sales. I would say that our image is steadily well-preserved. We are committed with our projects to sustain the business for a longer term. Besides, we still attend to complaints from our projects that have aged more than 20 years. This is considered that we are paying great attention on after-sale services.

It is a fact that PKNS has never abandoned a single project throughout its 52 years, and we are still maintaining the good record. So far, we have only one delayed projects and we manage to settle the problems promptly. We have introduced a new quality system to ensure a much higher and consistent standard of finishing our houses, with the setting up of PKNS Response Team to deliver maintenance services and also a customer service department. Meanwhile, for Datum series, if we put under the shoulder of PKNS, there will be a dilution effect or cross-poisoning effect since the market perception of other people is still ingrained with the existing image of PKNS which are famous in mass market and low-cost development. Due to that perspective, Datum is a brand by itself. Indeed, it is owned by PKNS, but the market segment is different. We do not want to jeopardise the image of both entities and poison each other among ourselves.

22.21 What Are the Other Factors That Have Made PKNS as the Preferred Housing Developers in Malaysia?

While awards may give a booster towards the good name of PKNS, the central point has always been about understanding the needs of our customers. One of the values here at PKNS is being 'customer-centric'. In order to live up to that value, we have the PKNS Careline, a combination of seven channels including a toll-free number, website, email and Facebook page where customers can send in their feedback, complaints and inquiries about PKNS. Moreover, PKNS is definitely not 'all talk and no action'. Once a call or report is received, PKNS will immediately respond to it, offering free maintenance and repair works to any of our strata-titled properties, including some of our flats that were built almost 40 years ago.

Throughout the years, PKNS has understood the need to strike a perfect balance of work, live and play. Truly, PKNS has come a long way to what it is today – a premier property developer that cares for the welfare of the underprivileged while introducing attractive and innovative products for all levels of society.

22.22 How Do You See the Pattern of Current Market Outlook in Selangor? Is There Any Momentous Influence of Environmental Forces, Such as Economic and Political Uncertainty Towards the Vision and Planning of PKNS?

The market outlook in terms of the offering is actually slowing down except for landed properties. In the near future, there will be a situation of oversupply for the commercial units that may create cautious sentiments among buyers for property market. Concomitantly with the uncertain scenario, PKNS is currently being circumspect regarding the market outlook for Malaysian property. Although some of our future developments are already in their pipeline, we have planned to curtail some of these investments and wait for the upside. PKNS as a property player in the market is keeping its radar on, but, regardless of obstacles, we will move forward and continue in our responsibility to develop properties for the people.

With respect to government regulation, houses that are priced between RM160,000 and RM250,000 or slightly higher to RM300,000 will encounter 70% of rejection rate on the borrowing. I perceive that this situation has to be relooked by the Central Bank of Malaysia (BNM), to encourage the population, especially the newly graduates and the newlyweds to buy their first house. I do understand that the data in regard to household debts is very scary, close to 88–90% our household debt to GDP ratio, but we have to find ways. This is something we are always in discussion with the state, to find ways how to be able to fulfil the needs. What we are afraid of, when the borrowing is too restrictive, then it is this market that will face the consequences, not the people who may have other options.

22.23 In Search for Excellence, What Is Your Personal Aim for PKNS?

My focal vision for PKNS is to create a sustainable organisation. I want PKNS to be a sustainable company, sustainable developer and sustainable group of organisations. Hence, my aim is to nurture a company that has good recurring income and be sustainable for at least another 50 years. If you look at historical data, close to about 90% of the companies were not able to sustain more than 50 or 60 years. A renewal process is hence needed for a company to remain competitive. You always need to do renewal – of the people, of the way you handle the business. By doing that, you can renew the organisation, and you can push forward. You must clear new ground, get new directions, achieve new targets and design it to go beyond certain terminal life. Holistically, we have made to sustain through high and low times, by delicately carving out blueprints for each and every township that we had built, making sure it is integrated with industrial components to create jobs, and then building enough houses and commercial offerings to support that new population.

22.24 So, What Are the Actual Targets of PKNS?

PKNS is transforming to be more effective and efficient, yet, the company will never forget its root. We are still going to strive forward. The make-up of our development will still be 85% low to affordable housing and middle-income housing. That is still our target. We will try to deliver the expectation of the public. There are also other property developers around, but they are mostly concentrating on the middle or upper classes of the public. Consequently, there is a need for 'knight developers' that will be able to reach into those urban poor or middle-income groups. This is where PKNS comes in.

22.25 As a Government-Linked Company (GLC), Do You Enjoy Any Special Advantages in Terms of Access to Funds, Tenders and Other Opportunities?

To be honest, no privileges have been given to us despite our status as one of the state-owned enterprises. In fact, PKNS does not receive grants from the government. Therefore, the company itself has to generate profits to subsidise for these homes and that is where the medium- to high-end developments step in. Profit is only a means to an end. The profits made are pumped back into the development of affordable homes as well as facilities and amenities for the community. In such essence, homebuyers should feel good because by buying a PKNS home, they are creating shelters for the underprivileged. While PKNS is diverse in its developments, one thing is for sure: buyers of any level are presented with a great investment opportunity. Generally, its products are priced 10–15% lower than others out on the market. Coupled with additional discounts for Bumiputera-allocated units, buyers are assured of valuable asset.

22.26 Do You Have Any Other Secret Mantra in Managing a Company?

I believe in the 'S curve'. When the curves are high, typically there are not many changes there, and everybody is very happy, but when you are declining, you need to put it to the right trajectory again. So, this is where transformation and reformation need to be done. With new and sophisticated challenges, PKNS must continue to evolve. We strive and practise strictest transparency and integrity in every dealing especially with other government agencies. We do not have certain privileges in terms of approval or other facilities, and we have to undergo similar processes, like other property developers.

22.27 Please Share with Us Your Chief Working Principles

Upon my graduation and since I am into the finance sector, my mother used to remind me to work wholeheartedly and honestly in every work. I have been trained since primary school to emulate my mother's attitude in budgeting the household's expenses. My father who was a headmaster had passed away since I was 4 years old, and my mother had to pawn her jewelleries to cover the cost of living. Fortunately, my mother is a resilient lady who is able to get back all the pawned items. In that sense, I would like to give credit to my beloved mother. Relatedly, along my way to this level, I have met with a number of great persons that have inspired me. They have taught me on the needs to be imbued with sharp business acumen and encouraged me to practise check and balance as well as not to trust others too much. The most important thing is you must have the correct support and people with you. We must work as a team. If we do not have expertise inside, we will bring them in. The organisational and financial structure must be simple and well-understood by people to reach certain objectives.

22.28 How Many Awards Have Been Obtained by PKNS Throughout the Lifespan of More Than 50 Years?

Strides made by PKNS in transforming its brand have not gone unnoticed as we have garnered international as well as local awards. Apart from GBI recognition, we have recently secured some awards on office development, office administration and Global Leadership Awards in 2013. In 2011, our project, Datum Jelatek, had received award of 'Most Iconic Design', while Asia Pacific Brands Foundation (APBF) has awarded The BrandLaureate – SMEs Best Brands (Corporate Branding) – SMEs Master Awards for two consecutive years, i.e. 2010 and 2011, respectively. After all, I would say that 2010 is the best year for PKNS. Despite the outstanding profit, Shah Alam was named as the 'Most Sustainable City' during the National Housing and Property Summit. Lastly, not to forget that we have been listed in Top 10 Property Developers at the BCI Asia Awards since 2012. More recently, we also receive 'Best Affordable Developer 2015' by Property Insight Award and were listed among the Top 10 Best Developers 2015 by i-Property.com.

Part V
Summary: Emerging Factors in Achieving Sustainable Competitive Advantage

Chapter 23
Conclusion

Keywords Summary and conclusion • Suggestion and recommendation • Internal and external factors • Business competitiveness • Malaysian housing developers

Viewed from the lens of housing developers, business success is realised once the houses are completed and handed over to proud home owners, preferably ahead of schedule. Concomitant with the points forwarded in the preceding chapters, both internal and external factors have been regarded as drivers towards the success of Malaysian private housing developers. Several personal characteristics of owner-managers have been revealed as well as key traits that contribute to business success. Successful housing developers remain consistently learnt and informed and are able to adapt to any regulatory changes or economic trends. Other attributes include hard work, patience and remaining focused. Overwhelming determination to succeed as well as an unyielding attitude to setbacks are vital ingredients in the extremely competitive real estate sector. In an industry known for constant challenges and unending predicaments, successful players exhibit tendencies of resilience, savvy and fortitude in the face of adversities; clearly this industry is not for the feeble-minded and weak of will.

In particular, housing developers need to possess doses of confidence, perhaps ruthless even, in grabbing business opportunities that are associated with a high degree of risks. Accordingly, potential housing developers in Malaysia should exude confidence, belief and conviction in their own abilities and that of their companies. The need to maximise profits should be coupled with the vision and ambition by implementing forward-thinking actions that will invariably enhance their sustainability in the industry. Undoubtedly, many developers are unable to anticipate turning points within the market signal, given the very volatile nature of the industry, so the need to be vigilant and perseverant is an added factor for success. Further, being the nexus within the urban development process, housing developers are required to exhibit the highest level of integrity and ethical probity in their project execution and be reflective of a working style synonymous with their final product that hopefully will generate 'a legion of followers'.

Based on the demographics, it can be revealed that an advantaged educational background would be an asset for housing entrepreneurs to achieve better performance in this overregulated industry. Although a number of senior successful play-

M. Jaafar et al., *Business Sustainability Model for Malaysian Housing Developers*, DOI 10.1007/978-981-10-5266-8_23

ers lack formal education, it would appear that for the forthcoming generation, a proper educational foundation is indispensable given the current dynamics of the industry which consistently deals with radical transformation. Similarly, experiential aspects will be added assets in enabling enhancement and development of entrepreneurial skills and abilities in order to adapt to change and confront unexpected challenges. Housing developers who enjoy similar industry experience might clearly benefit from easier transactions and better business performance.

A significant resource for housing developers to flourish in this industry concerns the land bank. As the issue of land scarcity becomes apparent, Malaysian housing developers need to be creative and innovative in dealing with land banking strategy. The senior management of development outfits requires astute and pertinent managerial capabilities in order to sustain their businesses and, in this context, should be imbued with the requisite skills of marketing, networking, planning and financial management. Businesses led by highly capable management translate to higher profits. Malaysian housing developers need to consistently apply and utilise novel marketing facilities to be competitive. The ability to recognise industry signals and analyse current market sentiments by implementing good networks with other stakeholders is a crucial asset. Pamadi, Ismael, and Haron (2015) postulate that housing developers should be imbued with enterprise agility and be able to develop their own network structure to acquire access to external resources in order to enhance their firms' performance.

In certain high-market growth areas, there is a need to have specific strategies to reflect the potential of rapid expansion plans. Almost every interviewed respondent covered in the book agrees on the importance of strategy and planning. The results indirectly contradict prevalent arguments espoused by Anikeeff and Sriram (2008) who claim many development companies lack long-term strategic plans. There is real value in daring to 'push the envelope' when it comes to business strategy, which will not only provide economic stability but impact upon firms' overall performance. In an industry well known for its high attrition rate, those able to keep costs manageable and spend prudently tended to reap the success. Still, what is clear is that in order to improve performance, housing firms need to keep restructuring the organisations and focus on strategy enhancement.

With respect to product characteristics, quality of the residential units on offer is perhaps the most critical facet to be considered since it deals with customers' satisfaction. Malaysian housing developers need to be committed to provide clients and potential purchasers with quality services in terms of cost-effectiveness as well as workmanship because buyers manifest interests only on the best overall product on offer. To warrant steady growth, developers need to take the initiative to be innovative towards the development processes by engaging in product innovation, as well as understand that uniqueness in the products' aesthetics and looks will inevitably influence customers' perception. From the findings, large property developers such as Ken Holdings Berhad and Darulaman Realty Sdn Bhd clearly highlighted on the importance of innovative housing concepts as unique features of their townships.

The twin aspects of project location and price play vital roles as well in determining the success of housing development. Most of the interviewees emphasised factors related to strategic locations, easy accessibility and comprehensive amenities as among the intangible factors that are perceived as desired qualities, all of which are considered value-added propositions towards attracting clients' interests. It is a truism to say that while price does matter, issues of quality and reputation actually find greater traction among buyers. Therefore, reputation and branding are somewhat intertwined and correlated, as well as imperative in order to attract desirable responses from buyers. Maintaining superior firm performance would thus require putting customers' needs as the top priority as it will definitely offer a positive image for future potentials. To outdo their rivals, developers need to put the highest emphasis on customer service; this goes beyond the projection of an excellent reputation but actually into bigger market share penetration that loyal and satisfied customers can be usually relied upon, without fail.

Consequently, housing developers may thus require undergoing the process of rebranding and 'administrative refurbishment'. Change though must be substantive, without which most rebranding exercise becomes the proverbial 'lipstick on a pig'. Clearly, superficial changes are inadequate to transform any organisation. Effective housing developers that are truly concerned about firm performance ought to pay attention to details, attentively listen to and respond to complaints and concerns expeditiously and represent their business in a positive and professional manner. Basically, housing developers should not just build homes but must try to look for ways to concertedly create better communities. Private housing developers in Malaysia, especially those with strong financial backups, must play their role as responsible corporate players by allocating a portion of their pecuniary success back to society by assisting the government to overcome problems on abandoned projects and build affordable homes for lower income citizenry. Through elements of corporate social responsibility (CSR), firms' project marketability and corporate reputation may be simultaneously enhanced and act as catalysts for ramping up firm performance.

Despite being receptive to new ideas, housing developers need to be responsive as well to the environment since the future of housing development is likely to be driven by the amalgamation of government policy on sustainability, the cyclical nature of boom-bust legacies as well as the rapid evolution of innovative technologies in the short and medium term. As with any industry, it is imperative to stay up-to-date with the laws, regulations, terminologies and trendsetting patterns that form the basis of the real estate business. Housing developers need to always remain calm, in the inevitable turbulence that characterises the industry. While the legacy of cyclical economic setbacks remain stark and real, growing homebuilders need to navigate and manage expansion into new markets carefully and intelligently, without the pressure of increasing overheads or retraining the entire staff, in essence, to prudently administer business costs at all times. Perhaps, without sounding pessimistic, housing developers need to always consider the worst-case scenario and strategically decide the next growth trajectory, which may or may not be, in real estate. In any case, reality demands that an exit strategy be emplaced.

Generally, the current state of the economy as well as government initiatives will have an impact on the performance of housing developers due to the legislative roles of government and policies that affect macroeconomic stability. Developers are advised to stay nimble, conduct necessary structural adjustments and be adaptive and flexible to changes during tough times of economic downturns and in the absence of government intervention measures. Prudent and experienced housing developers understand the risks – not only in terms of real estate deals but also the legal implications involved – and adjust their businesses to mitigate those risks. The sequential nature of the housing development process, which covers the moment projects are conceived to the physical construction until the product delivery, enables sharp housing developers to make quick adjustments to their overall strategy enabling, at the very least, to minimise the potential of imminent huge losses. Apart from that, since the majority of Malaysian housing developers rely upon external funds, a comprehensive and enabling internal financial management system should be incorporated for regular monitoring and requisite periodic audits, all to ensure that the project capital is ring-fenced against external 'shocks' and unforeseen circumstances.

Successful developers are also often engaging and tuned in with the latest trends and customer preferences. The need to invest in a significant amount of time on every element of the homebuilding process, from selecting the finest quality materials, exploring new design possibilities to instilling pride to every member of the design, construction and sales team, can only translate into improved financial performance. From the human resource management perspective, developers should assemble a team of dedicated and talented people and to proactively and skilfully lead them for the best possible outcome; in other words, the team must make things happen. Owner-managers should also try to encourage a corporate culture that engenders employee loyalty since great companies are derived from the industry and focus of its people, who are uncompromising and undeterred in the face of setbacks and failures. Another truism of the housing development industry is this – loyal staffs that are sales-oriented, competent, diligent, results-oriented and service-oriented are the greatest assets of the business. Developers need to move from mere people selection to proper talent identification of human resource potential that can be nurtured and imbued with the culture of excellence and belonging; while talent identification should be carefully done and performance monitored consistently, management has to recognise the value of proper and periodic training as well.

Accordingly, it is suggested that performance-oriented housing developers should be transparent in terms of employee performance and set clear incentives and rewards system. As mentioned, there is a need to provide a real sense of belonging by putting in place opportunities for career advancement within the company, offering a clear understanding of business priorities with attractive remuneration packages, providing a dynamic and exciting place to work and promoting an open communication culture that encourages direct contact with all levels of management. These are but some of the battery of issues that Malaysian private housing developers are well advised to heed for consolidation and growth. Behavioural alignment is therefore viewed as sine qua non for organisations to operate successfully.

The real success of housing provision can be further ensured by leadership that entails good and responsible governance through careful maximisation of resources by synthesising all the foregoing success factors. Personal missions and visions need to be clarified and be in tandem with the greater organisational good in order to establish a strong presence in the property sector. Business objectives can be fulfilled if they can deliver long-term value to stakeholders by controlled expansion balanced with aggressive competition. Although success in housing development is highly dependent on the developers' composition, land location and development planning, homebuyers' welfare constitutes another instrumental feature for their success. Malaysian housing developers should therefore be knowledgeable – both in terms of industrial and managerial knowledge because knowledge has been recognised as a source of competitive advantage particularly during times of challenging environment, uncertainty and long-term business performance outlook.

Housing provision is an issue that looks likely to continue to vex the government. The government faces the complex dilemma of having to take care of the welfare of the public as well as those of the developers. That being the case, both state governments and housing developers need to work closely at all times to not only meet the state's housing needs but also to enable the property industry to contribute to the state's growth. But political exigencies and commercial priorities do not always make good alliances. Given the already inherently competitive nature of the industry and limited land banks, private developers face a huge conundrum in the zero-sum scheme of things. Developers are reminded however to be resourceful and resilient in the face of massive obstacles, none more so than those from the state authorities. But experience will always remain the best guide, for without proper guidance, the nature of the housing business is too perilous to warrant involvement. In the present climate of political turmoil and economic uncertainties, both firms and owners should stay true to the twin tenets of being agile and flexible to integrate into the dynamics of this volatile industry.

The current business setting, uncertain at best and turbulent at worst, has also not spared housing developers in Malaysia; performances have been marred by a number of challenges, mostly emanating from government-driven policies. In order to overcome such hurdles, it is suggested that relevant policies be reviewed and measures revised to enable a uniform mutually beneficial position to be adopted by all stakeholders. Among the notable challenges extracted from the interviews were the issues related to securing of bank loans – the tedious and complicated processes have frustrated a lot of potential buyers, thereby impacting possible sales. Terms and conditions are getting stricter, and a plethora of 'absurd requirements' is inserted to deny potential buyers of housing financing. It becomes imperative for developers therefore to have adequate financial backups and spare capital, as the upfront financial commitment in this sector is high and draining on cash reserves. It was also gleaned from the interviews about the unpredictability and untimeliness regarding the government's role and intervention, which has only led to more confusion. The rules and policies promulgated by the government tend to change profusely without considering the developers' hardships. Although the intention was to reduce house prices, thereby making houses more affordable to the citizenry, profit margins of private housing developers are subsequently diminished. Moreover, with skyrocketing land prices each year that have a direct influence on the price of houses, the 'worst-case scenario' of financial ruin could thus translate into stark reality. Although housing demand is still being maintained, the scenario will change dramatically if the price speculation is not curbed, as more and more developers adopt a 'wait-and-see' attitude.

Another big regulatory issue is the minimum of 30% of private sector mixed development required to be set aside for low-cost housing that was introduced as a precondition for planning approval. The issue of housing developers cross-subsidising low-cost housing with high-end projects has been much discussed without any clear position as to its finality, essentially due to the intrinsic differences in costs of land related to project locality, as well costs of construction due to project types. It is clear that the entire matter has no easy or quick resolution, no matter how well-intentioned the policy was intended to be. Perhaps public developers could supplant or take the responsibility to build affordable houses for nonaffluent citizens, or otherwise the government could offer incentives and proactive measures to undertake and construct low-cost units, or build them on time. There have also been complaints towards the local authorities on the time taken to endorse land-use approvals (Abdul-Aziz, Ho, & Jaafar, 2006), with such delays consequently leading to project failure. These predicaments are further exacerbated, since falling under the jurisdiction of each state government, these guidelines tend to vary from one state to another. This clearly illustrates the need for housing developers to build strong networking relationships with local agencies to expedite the process.

To add to the litany of current challenges mentioned is the issue associated with escalating costs, aggravated by increasing prices of building materials, the weakening Ringgit and the imposition of the Goods and Services Tax (GST). The increasing cost of raw materials and the imposition of minimum wages directly affect the ability of housing developers to hold down the price of houses. A number of housing developers lamented on the costs of doing business that have gone up severely – rising land conversion costs along with capital contribution and compliance costs – makes the profit proposition even more difficult with slim and, at times, untenable margins. Therefore, the need for accuracy in project cost computation is another paramount and essential feature for successful developers; essentially, the price and quantity of housing to be built are dependent on the price and quantity of the factors of production. This covers the basic economic concept of the theory of supply and demand, and it is a given that the private sector is more concerned towards profit, while the government sector is more inclined towards providing for the welfare of the people. In other words, Malaysian housing developers need to have clarity of purpose in their corporate mission and programme their business undertakings in accordance with the current environment. Importantly, they should also realise that they are the prime mover and also the backbone of national development in Malaysia; monetary gains should not always be the overriding factor, as affordable housing issues require addressing as well – striking the right balance would be the crucial and responsible response.

Malaysian housing developers, to put it bluntly, should not be burdened by a stringent loan policy and overbearing government bureaucracy. A nationally promulgated housing policy may produce different housing outcomes at the regional level not only because of regional variations in the housing market but also due to the country's legal and governance system. The government should exercise a more pragmatic approach by collaborating with private housing developers and re-examine the existing rules and regulations that proffer win-win situation to all stakeholders. The establishment of the National Housing Council to develop strategies and action plans in a holistic manner in ensuring more efficient and expeditious housing delivery is timely indeed. The government should also be ever ready to revise any policies related to housing development and encourage the financial institutions to revise their rules and procedures relating to the approval of bridging and home financing loans, in particular one which relates to foreign purchasers. Current policy making initiatives have to take cognisance of both specific national, economic and political factors and wider forces at work in the global economy.

The book has listed the success factors above, but mentioned the need for caution and care in the obviously foreboding nature of the real estate sector. This industry is not for the faint-hearted nor for the gung-ho maverick. The wide array of success factors would just be as applicable to other industries as they are to housing development, although the specific nature of the industry – land, local authorities, professionals, marketing, time, etc. – makes for a rather unique and challenging setting. Potential new entrants must learn from the experiences and mistakes of those whom the book has covered fairly exhaustively. The book also highlights the need for the government, property players, professional bodies and academia to work together to revisit existing standards, rules and laws by incorporating contemporary and sustainable needs. As Malaysia aspires to become a fully developed country by 2020, the valuable knowledge obtained through academic research should be effectively disseminated to the industry to fundamentally reduce the gap between theory and practice. It is postulated that new market entrants can utilise the findings to amass the necessary resources for effective market positioning, while other stakeholders can utilise them as guidelines to indirectly reduce occurrences of recalcitrant and unscrupulous developers. Ultimately, it is the homeowner that benefits.

References

Abdul-Aziz, A. R., Ho, S. Y., & Jaafar, M. (2006). Competitive resources of private housing developers: The Malaysian perspective. *Journal of Engineering, Design and Technology, 4*(1), 71–80.

Anikeeff, M. A., & Sriram, V. (2008). Construction management strategy and developer performance. *Engineering, Construction and Architectural Management, 15*(6), 504–513.

Pamadi, M., Ismael, I., & Haron, H. (2015). The effect of enterprise agility and network structure on firms' performance in the housing industry. *Advanced Science Letters, 21*(4), 643–647.

Index

M. Jaafar et al., *Business Sustainability Model for Malaysian Housing Developers*, DOI 10.1007/978-981-10-5266-8

V

W

Z

MIX
Papier aus verantwortungsvollen Quellen
Paper from responsible sources
FSC® C105338

If you have any concerns about our products,
you can contact us on
ProductSafety@springernature.com

In case Publisher is established outside the EU,
the EU authorized representative is:
Springer Nature Customer Service Center GmbH
Europaplatz 3, 69115 Heidelberg, Germany

Printed by Libri Plureos GmbH
in Hamburg, Germany